T0296138

Cambridge Tracts in Mathematics
and Mathematical Physics

GENERAL EDITORS
G. H. HARDY, M.A., F.R.S.
E. CUNNINGHAM, M.A.

No. 34

THE RATIONAL QUARTIC CURVE

THE
RATIONAL QUARTIC CURVE

IN SPACE OF THREE AND FOUR
DIMENSIONS

Being an Introduction
to
Rational Curves

by

H. G. TELLING

CAMBRIDGE
AT THE UNIVERSITY PRESS
1936

CAMBRIDGE
UNIVERSITY PRESS

University Printing House, Cambridge CB2 8BS, United Kingdom

Cambridge University Press is part of the University of Cambridge.

It furthers the University's mission by disseminating knowledge in the pursuit of education, learning and research at the highest international levels of excellence.

www.cambridge.org
Information on this title: www.cambridge.org/9781107493964

© Cambridge University Press 1936

First published 1936
Re-issued 2015

A catalogue record for this publication is available from the British Library

ISBN 978-1-107-49396-4 Paperback

CONTENTS

CHAPTER II

RATIONAL QUARTIC CURVE C_1 IN [3]

NOTE ON INVOLUTIONS ON C

PREFACE

This essay, first drafted as a Fellowship Essay, was originally much longer. The necessary compression has involved that some of the subsidiary material should be left in the form of exercise or suggestions for the reader, and it is hoped that this arrangement may have the advantage of emphasising the main line of development.

References to the literature are so fully and compactly given in the Encyclopaedia articles, mentioned on p. viii, that it has seemed unnecessary to repeat them here, except in the case of particular points on which a reader may be likely to demand further information.

My thanks are due to Professor Baker without whose constant help and encouragement the essay would never have been written.

To the Press I am indebted for their courtesy; the reader will doubtless join me in appreciation of their elegant handling of the formulae.

H. G. T.

NEWNHAM COLLEGE
    CAMBRIDGE

BIBLIOGRAPHY

Encycl. der Math. Wiss. III, C. 7, C. Segre—especially § 22 and § 27.

III, C. 9, K. Rohn u. L. Berzolari, §§ 57–60.

III, C. 10, W. Fr. Meyer, §§ 42–51.

In these articles detailed references are given.

Pascal's *Repertorium der höheren Math.* II, 2 (1922), Kap. XXIX C, XXXVII.

Reference may also be made to the following books:

E. Bertini. *Introduzione alla Geometria proiettiva degli iperspazi.*

E. Ciani. *Introduzione alla Geometria algebrica.*

J. Fr. De Vries. *Analytische behandeling van de rationale kromme van den vierden graad in een vierdimensionale ruimte.*

F. Enriques e O. Chisini. *Teoria geometrica delle equazioni*, T_3.

J. H. Grace and A. Young. *Algebra of Invariants.*

G. Loria. *Curve sghembe speciali algebriche e trascendenti*, T_1.

W. F. Meyer. *Apolarität u. rationale Kurven.*

G. Salmon. *Geometry of Three Dimensions.*

For historical sketch with references see:

H. W. Richmond. *Camb. Phil. Trans.* 19 (1900) 132.

CHAPTER I

RATIONAL QUARTIC CURVE C IN [4]

Linear transformations

1·1. If x_i, y_i $(i = 0, 1, 2, 3, 4)$ are the projective coordinates of points (x), (y) in four-dimensional spaces X, Y, the linear transformation

$$x_i = \sum_k a_{ik} y_k \quad (i, k = 0, 1, 2, 3, 4),$$

where the determinant of the coefficients $|a_{ik}|$ does not vanish, is interpreted geometrically as a collineation, i.e. a one-one correspondence between the points such that the points of a line, plane, prime of X correspond to points of a line, plane, prime of Y. There are twenty-four essential constants in the transformation; it is possible to determine a collineation (uniquely) so that six given points of X transform into six given points of Y, each five of the six points in both cases being linearly independent but otherwise arbitrary. If the six points of the reference system in X correspond to the points of the reference system in Y, the collineation is simply $x_i = y_i$. Hence the general linear transformation above may equally well be regarded as the change of reference system in X, where x_i and y_i are now the old and new coordinates of the same point of X.

If the spaces X and Y are superposed there are in general five linearly independent points which are transformed into themselves. These are called the double points of the collineation. Referred to the pentad of double points and any point as unit point, the equations of the collineation become

$$x_0 : x_1 : x_2 : x_3 : x_4 = \rho_0 y_0 : \rho_1 y_1 : \rho_2 y_2 : \rho_3 y_3 : \rho_4 y_4; \quad 1·11$$

and the ratios $\rho_0 : \rho_1 : \rho_2 : \rho_3 : \rho_4$, which are the coordinates of the point corresponding to the unit point, are called the invariants of the collineation.

Particular cases of special importance arise when in 1·11:

$$\text{(i)} \quad \rho_0 = \rho_1 = \rho_2 = 1, \quad \rho_3 = \rho_4 = \rho,$$

in which case every point of the line $x_0 = x_1 = x_2 = 0$, and every point of the plane $x_3 = x_4 = 0$ are double points. The line joining any two corresponding points (x) and (y) meets this line and plane; and ρ, the one invariant, is the cross-ratio of the four points thus obtained upon the joining line. In the case when $\rho = -1$, this collineation is called *harmonic inversion*. If the line and plane of double points are given, the point (y) corresponding in the harmonic inversion to any point (x) may be obtained by drawing through (x) the unique line which meets the line and plane of double points, and taking (y) on this line such that (x) and (y) are harmonically separated by the line and plane.

$$\text{(ii)} \quad \rho_0 = \rho_1 = \rho_2 = \rho_3 = 1, \quad \rho_4 = \rho,$$

in which case every point of the prime $x_4 = 0$ and the point $(0, 0, 0, 0, 1)$ are double points. This is the case of perspective: the line joining (x) and (y) passes through a fixed point, the isolated double point, O; the four points, (x), (y), O and the intersection of the line with $x_4 = 0$, have cross-ratio ρ. When $\rho = -1$, the collineation is called *harmonic perspective*.

It should be noted that if in 1·11 the collineation is involutory, so that to (x) corresponds (y) and in the same sense to (y) corresponds (x), we must have

$$\rho_0^2 = \rho_1^2 = \rho_2^2 = \rho_3^2 = \rho_4^2 = 1,$$

and the only cases which arise apart from identical transformation are the harmonic cases above.

Definition of C

2·1. A curve is said to be *rational* when the coordinates of its points can be expressed as rational functions of one parameter. The parameter can always be chosen so that to each value of the parameter corresponds one and only one point of the curve,* i.e. there is a one-one algebraic correspondence between the points of the curve and the points of a line.

Any algebraic curve of order four lies in a space of n dimensions, $n \leqslant 4$, for if the curve were in [5] a prime could be drawn through five general points of the curve and would then contain it entirely.

* Lüroth, *Math. Ann.* **9** (1876) 163.

Every algebraic curve of order four in [4] is rational and without singularities, its points being in one-one correspondence with the points of a line: for a pencil of primes may be drawn through three general points of the curve, each prime having one further intersection with the curve and a corresponding point of intersection with the line; moreover, if the curve has a node, the prime determined by any three points of the curve together with the node has five intersections with the curve and therefore contains it.

The rational quartic curve in [4], C, is given by

$$x_0 : x_1 : x_2 : x_3 : x_4 = f_0(t) : f_1(t) : f_2(t) : f_3(t) : f_4(t),$$

where $f_0, \ldots, f_4$ are polynomials in t, and since the curve is of order four it is met by an arbitrary prime $\Sigma \xi_i x_i = 0$, in four points whose parameters are the roots of $\Sigma \xi_i f_i = 0$, so that the polynomials must be such that at least one is of the fourth order in the parameter t, and none is of higher order. Thus the coordinates of a point (x) of C are given by

$$x_i = \sum_{k=0}^{4} a_{ik} t^k \quad (i = 0, \ldots, 4),$$

wherein at least one of the coefficients a_{i4} is not zero; further, the condition $|a_{ik}| \neq 0$ is necessary to ensure that the curve C does not lie in a space of less than four dimensions.

If $|a_{ik}| = 0$ but the matrix of coefficients is of rank four, then the polynomials $f_i(t)$ are linearly dependent and there is a single identical relation $\Sigma \lambda_i x_i = 0$, which is the equation to a prime in which the curve lies. This prime may then be taken as the fifth prime of the pentad of reference, and the curve lying therein is referred to the tetrad $x_0 = 0$, $x_1 = 0$, $x_2 = 0$, $x_3 = 0$, of planes in which the other primes of the pentad meet the fifth. The curve then has equations

$$x_i = \sum_{k=0}^{4} a_{ik} t^k \quad (i = 0, \ldots, 3),$$

in the prime in which it lies. This curve, which is the general rational quartic curve in [3], may be considered as the projection from the point $(0, 0, 0, 0, 1)$ in [4] upon the prime $x_4' = 0$ of any rational quartic curve in [4] which is given by the above equations together with $x_4' = \Sigma a_{4k}' t^k$, where the coefficients a_{4k}' are arbitrary subject to the non-vanishing of the new determinant of the coefficients.

Further, if the matrix of the coefficients in the equations of C is of rank three, there are two linear identities in the polynomials and the curve lies in a plane. If the first three polynomials are those which are linearly independent, the curve when referred to a system of coordinates in the plane in which it lies has equations

$$x_i = \sum_{k=0}^{4} a_{ik} t^k \quad (i = 0, 1, 2).$$

This curve, the general rational quartic in the plane, may be considered as the projection of a rational quartic in [4] given by the above equations together with $x_i' = \Sigma a_{ik}' t^k$ $(i = 3, 4)$, the projection being from the line $x_0 = x_1 = x_2 = 0$ upon the plane $x_3' = x_4' = 0$.

The rational quartic curve in [4] is *normal* in that it is not the projection of any quartic curve in space of higher dimensions. *Every irreducible rational quartic curve which is not normal is a projection of the normal curve.*

Osculating system of reference

2·2. The equations of the curve C,

$$y_i = \Sigma a_{ik} t^k, \quad |a_{ik}| \neq 0,$$

are reduced by the linear transformation

$$y_i = \Sigma (-1)^k a_{ik} x_k$$

to the form
$$x_i = (-1)^i t^i.$$

The linear transformation is equivalent to the choice of an *osculating system of reference*, and we proceed to find the relation of such a system to the curve. The pentad of reference is $A_0, \ldots, A_4$, where $A_0 = (1, 0, 0, 0, 0)$, etc.; A_0, A_4 and the unit point are on the curve at the points whose parameters are $0, \infty, -1$ respectively. At A_0 $(t = 0)$ we have

the osculating prime $A_0 A_1 A_2 A_3 : x_4 = 0,$

the osculating plane $A_0 A_1 A_2 \quad\quad : x_3 = x_4 = 0,$

the tangent $A_0 A_1 \quad\quad\quad\quad : x_2 = x_3 = x_4 = 0;$

and similarly at A_4 $(t = \infty)$. The point A_2 is that common to the osculating planes at A_0 and A_4, and $A_0 A_1 A_3 A_4$ ($x_2 = 0$) is the prime bitangent to C at A_0 and A_4; while $A_0 A_4$ ($x_1 = x_2 = x_3 = 0$) is the chord joining A_0 and A_4, and $A_1 A_2 A_3$ ($x_0 = x_4 = 0$) is the axis plane common to the two osculating primes.

A linear transformation of the parameter, $\bar{t} = (at + b)/(ct + d)$, admits of the choice of $a : b : c : d$, and thus any three points of the curve can be given the parameters 0, ∞, -1. By taking the first two of these as A_0, A_4, an osculating pentad of reference can be constructed; the third point is taken as unit point. Hence there are ∞^3 osculating systems of reference; for each of these the co-ordinates of the curve are given as $x_i = (-t)^i$, and to pass from one to the other is equivalent to a linear transformation of t and a consequent linear transformation of the coordinates (x).

Projective generation of C

2·3. From the equations $x_i = (-t)^i$, it follows that C lies on six linearly independent quadrics whose equations are the determinants of the array

$$\begin{Vmatrix} x_0 & x_1 & x_2 & x_3 \\ x_1 & x_2 & x_3 & x_4 \end{Vmatrix} = 0. \qquad 2·31$$

In [4] there are in all ∞^{14} quadrics; to pass through C imposes on a quadric nine linear conditions, so that the complete system of quadrics through C is a linear combination of those given by 2·31.

The array suggests two complementary generations of the curve C. On the one hand the curve is the locus of the point of intersection of the related pencils of primes

$$x_0 + t x_1 = 0, \quad x_1 + t x_2 = 0, \quad x_2 + t x_3 = 0, \quad x_3 + t x_4 = 0.$$

On the other hand corresponding primes of the two projective stars of primes, through the points $(0, 0, 0, 0, 1)$ and $(1, 0, 0, 0, 0)$, given by

$$\left. \begin{aligned} \lambda_0 x_0 + \lambda_1 x_1 + \lambda_2 x_2 + \lambda_3 x_3 = 0, \\ \lambda_0 x_1 + \lambda_1 x_2 + \lambda_2 x_3 + \lambda_3 x_4 = 0, \end{aligned} \right\} \qquad 2·32$$

intersect in planes; then (x) being given subject to 2·31 we have ∞^2 (and thus three linearly independent) solutions for $\lambda_0 : \lambda_1 : \lambda_2 : \lambda_3$; thus of the one star the primes so obtained meet in a line, and the corresponding primes of the other star meet in the corresponding line; and these two lines meet in the point (x) of C. Thus the quartic curve is the locus of points of intersection of corresponding lines of two related stars of primes.

These complementary methods of generation arise in a slightly more general form from the array

$$\left\| \begin{array}{cccc} y_1 & y_2 & y_3 & y_4 \\ z_1 & z_2 & z_3 & z_4 \end{array} \right\| = 0, \qquad 2\cdot33$$

where y_i, z_i are any linear functions of the coordinates. In this case the base planes, $y_i = z_i$, of the pencils are four trisecant planes of C, while $y_1 = y_2 = y_3 = y_4 = 0$ and $z_1 = z_2 = z_3 = z_4 = 0$ are two points on C which are the base points of the stars.

A more descriptive method of exhibiting the generation by projective stars is obtained by taking a section of the stars by a general prime. In the section the corresponding lines, planes and primes of the stars appear as corresponding points, lines and planes of a collineation. There are in general four double points of this collineation and they are given by corresponding lines of the stars which meet in the section. Thus the locus of the intersection of corresponding lines is a quartic curve through the base points of the stars. The corresponding planes of the stars are determined by pairs of corresponding lines and thus meet in chords of the quartic curve; the corresponding primes are determined by triads of corresponding lines and hence meet in planes each containing three points of the curve.

A quartic curve can be described through seven general points A_i, by either of the foregoing complementary methods of generating the curve. With A_1 and A_2 as centres, the five lines $A_1 A_3$, $A_1 A_4$, ..., $A_1 A_7$, corresponding to the lines $A_2 A_3$, $A_2 A_4$, ..., $A_2 A_7$, determine related stars whose corresponding lines intersect in points of the curve. On the other hand four projective pencils of primes with bases $A_2 A_3 A_4$, $A_3 A_1 A_4$, $A_1 A_2 A_4$, $A_1 A_2 A_3$, are determined by three corresponding sets given by the primes to A_5, A_6, A_7, respectively, and the intersection of these primes generates a quartic curve through these three points and A_1, A_2,

A_3, A_4. *Thus seven general points determine uniquely a quartic curve through them.*

2·3,1. A unique quartic curve can be constructed having given r trisecant planes and $7-r$ points of the curve, in the cases when $r = 0, 1, 2, 3, 4, 5$.*

2. The trisecant planes of C meet a general fixed prime in the lines of a tetrahedral complex.

3. The cross-ratio of the four primes joining a variable point on C to four fixed trisecant planes is constant.

4. A variable tangent of C meets four fixed osculating primes in a constant cross-ratio.

5. Corresponding lines of a collineation in [3] which meet do so on a four-nodal cubic surface.

6. If ten points on C be 1, 2, 3, 4, 5, 6, 7, 8, 9, 0, the primes 1234 and 6789, 2345 and 7890, 3456 and 8901, 4567 and 9012, 5678 and 0123, meet in five planes which are associated.†

7. If ten points are the intersections of a quadric with an elliptic quintic curve in [4], the quartic curve determined by seven of them meets the plane through the remaining three in three points.

8. If $P_i = 0, (i = 1, ..., 10)$, are the equations of ten points of C, then there is a relation $\sum_1^{10} P_i^2 \equiv 0$.

If ten points in [4] are related by $\sum_1^{10} P_i^2 \equiv 0$, the quartic curve through seven of the points has the plane containing the other three as trisecant plane. The proposition 2·3, 6 holds for any ten points for which $\sum_1^{10} P_i^2 \equiv 0$.‡

9. Quartic curves in [4] which pass through six fixed points meet a general prime in tetrads which are polar with respect to a fixed quadric Q in that prime.

The prime is met by the lines, planes and primes joining the six points in the configuration arising from two tetrahedra in perspective; this figure is self-reciprocal with respect to Q.

---

* White, *Jour. L.M.S.* 4 (1929) 11. The result is also true when $r = 6$; Todd, *Proc. Camb. Phil. Soc.* 26 (1930) 323; for $r = 7$ there are ten curves; Babbage, *Jour. L.M.S.* 8 (1933) 9.

† Cf. Weddle's theorem in Reye, *Geometrie der Lage*, 3 (1923) 229.

‡ Cf. P. Serret, *Géométrie de direction* (1869) 311.

10. If (x) and (y) are two points on the same quartic curve through the six reference points in [4], then t and θ can be found such that

$$t/x_i + \theta/y_i = 1 \quad (i = 0, 1, ..., 4).$$

If (x) and (y) are in the prime $\Sigma \xi_i x_i = 0$, they are conjugate with respect to the quadric $\Sigma \xi_i x_i^2 = 0 = \Sigma \xi_i x_i$.

Fundamental polarity

2·4. In what follows the equations of the curve C are taken as $x_i = (-t)^i$ and a point whose parameter on C is t is spoken of as the point t.

A prime (ξ) meets C at points given by the roots of the equation

$$\xi_0 - \xi_1 t + \xi_2 t^2 - \xi_3 t^3 + \xi_4 t^4 = 0.$$

The coordinates of the prime are thus proportional to the elementary symmetric functions of these roots, so that if t_1, t_2, t_3, t_4 are four points of the curve, the equation of the prime containing them is

$$x_0 t_1 t_2 t_3 t_4 + x_1 \Sigma t_1 t_2 t_3 + x_2 \Sigma t_1 t_2 + x_3 \Sigma t_1 + x_4 = 0. \qquad 2\cdot41$$

If $t_1 = t_2 = t_3 = t_4 = t$, this prime is the osculating prime at t; its equation is

$$(xt)^4 \equiv x_0 t^4 + 4x_1 t^3 + 6x_2 t^2 + 4x_3 t + x_4 = 0, \qquad 2\cdot42$$

and its coordinates are

$$\xi_0 : \xi_1 : \xi_2 : \xi_3 : \xi_4 = t^4 : 4t^3 : 6t^2 : 4t : 1.$$

There are therefore *four osculating primes through a given point* (x), their points of contact being the roots of the equation 2·42. The prime (ξ) determined by the points of contact is called the *polar* prime of (x) and (x) the *pole* of (ξ), the formulae connecting (x) and (ξ) being

$$\xi_0 = x_4, \quad \xi_1 = -4x_3, \quad \xi_2 = 6x_2, \quad \xi_3 = -4x_1, \quad \xi_4 = x_0.$$

Two points (x) and (y) are conjugate in the polarity when

$$x_0 y_4 - 4x_1 y_3 + 6x_2 y_2 - 4x_3 y_1 + x_4 y_0 = 0;$$

and pole and polar prime are incident for points and primes of the quadric

$$i \equiv x_0 x_4 - 4x_1 x_3 + 3x_2^2 = 0. \qquad 2\cdot43$$

This polarity is fundamental in the geometry of C; we shall refer to the quadric which is the nucleus of the polarity as I.

In consequence of the polarity the complete figure of C consisting of its points, tangents, osculating planes and osculating primes is self-dual; to a chord corresponds an axis plane common to two osculating primes, to a trisecant plane an axis line, and to a bitangent prime a point common to two osculating planes.

The polar prime of any point of C is the osculating prime at the point and the tangent prime of I. The polar plane of a tangent of C is that common to the two consecutive osculating primes at the two consecutive points which determine the tangent, and is thus the osculating plane at the point of contact of the tangent. As the osculating plane contains the tangent, it follows that any tangent of C lies on I, and this may be verified by direct substitution, the points of the tangent at t being given by

$$(1, \ -t-\lambda, \ t^2+2\lambda t, \ -t^3-3\lambda t^2, \ t^4+4\lambda t^3).$$

Thus as a locus the quadric I contains the curve C and its tangents; and as an envelope contains the osculating primes and osculating planes of C.

Trisecant planes

2·5. A plane containing three points t_1, t_2, t_3 lies in a prime with any fourth point t_4. If (x) is a point on this plane, the formula 2·41 holds for any value of t_4. Hence the equations of the trisecant plane are

$$\left. \begin{aligned} x_0 t_1 t_2 t_3 + x_1 \Sigma t_1 t_2 + x_2 \Sigma t_1 + x_3 = 0, \\ x_1 t_1 t_2 t_3 + x_2 \Sigma t_1 t_2 + x_3 \Sigma t_1 + x_4 = 0, \end{aligned} \right\} \qquad 2\text{·}51$$

where the summations are taken over t_1, t_2, t_3, or

$$\lambda_0 x_0 + \lambda_1 x_1 + \lambda_2 x_2 + \lambda_3 x_3 = 0,$$
$$\lambda_0 x_1 + \lambda_1 x_2 + \lambda_2 x_3 + \lambda_3 x_4 = 0$$

(as in 2·32), and the points of intersection t_1, t_2, t_3 with the curve are the roots of
$$\lambda_0 - \lambda_1 t + \lambda_2 t^2 - \lambda_3 t^3 = 0.$$

The two equations are linearly independent except when

$$\begin{Vmatrix} x_0 & x_1 & x_2 & x_3 \\ x_1 & x_2 & x_3 & x_4 \end{Vmatrix} = 0;$$

consequently there are ∞^1 trisecant planes through any point (x)

which is not on C. If a trisecant plane passes also through a second point (y), then (x) and (y) are connected by the relation

$$\begin{vmatrix} x_0 & x_1 & x_2 & x_3 \\ x_1 & x_2 & x_3 & x_4 \\ y_0 & y_1 & y_2 & y_3 \\ y_1 & y_2 & y_3 & y_4 \end{vmatrix} = 0, \qquad\qquad 2\cdot 52$$

which is therefore the equation to the locus generated by the trisecant planes through a fixed point (y). Thus *the ∞^1 trisecant planes through a point generate a quadric cone.*

The equations of the osculating plane at t are obtained by putting $t_1 = t_2 = t_3 = t$ in 2·51 or by regarding the plane as the intersection of consecutive osculating primes; they are thus

$$\left.\begin{aligned} x_0 t^3 + 3x_1 t^2 + 3x_2 t + x_3 = 0, \\ x_1 t^3 + 3x_2 t^2 + 3x_3 t + x_4 = 0. \end{aligned}\right\} \qquad 2\cdot 53$$

The elimination of t between these two equations shews the locus of osculating planes to be a sextic primal; thus six osculating planes meet a general line.

Chords

2·6. A chord joining the points t_1, t_2 lies in the same prime with any other two points t_3, t_4; thus from 2·41, any point (x) on the chord satisfies the equations

$$\left.\begin{aligned} x_0 t_1 t_2 + x_1 (t_1 + t_2) + x_2 = 0, \\ x_1 t_1 t_2 + x_2 (t_1 + t_2) + x_3 = 0, \\ x_2 t_1 t_2 + x_3 (t_1 + t_2) + x_4 = 0, \end{aligned}\right\} \qquad 2\cdot 61$$

or
$$\begin{aligned} \lambda_0 x_0 + \lambda_1 x_1 + \lambda_2 x_2 = 0, \\ \lambda_0 x_1 + \lambda_1 x_2 + \lambda_2 x_3 = 0, \\ \lambda_0 x_2 + \lambda_1 x_3 + \lambda_2 x_4 = 0, \end{aligned}$$

its intersections t_1, t_2 with the curve being given by
$$\lambda_0 - \lambda_1 t + \lambda_2 t^2 = 0.$$

Thus the chords of C generate the cubic primal, J, whose equation is

$$j \equiv \begin{vmatrix} x_0 & x_1 & x_2 \\ x_1 & x_2 & x_3 \\ x_2 & x_3 & x_4 \end{vmatrix} = 0; \qquad\qquad 2\cdot 62$$

and the primal appears as the locus of the lines of intersection of corresponding members of three projective nets of primes. The base lines of the nets are the tangents at A_0 and A_4 and the chord A_0A_4. But as the equation 2·62 may be written

$$\begin{vmatrix} u_0 & u_1 & u_2 \\ v_0 & v_1 & v_2 \\ w_0 & w_1 & w_2 \end{vmatrix} = 0,$$

where u_i, v_i, w_i are the same linear functions of $x_0, x_1, x_2; x_1, x_2, x_3; x_2, x_3, x_4$, the primal may be regarded as generated by projective nets with any three chords of C as base lines.

The tangents may be considered as limiting positions of chords and thus also lie on J. By putting $t_1 = t_2 = t$ in 2·61, or by regarding the tangent as the intersection of three consecutive osculating primes, the equations of the tangent at t are obtained as

$$\left. \begin{aligned} x_0 t^2 + 2x_1 t + x_2 &= 0, \\ x_1 t^2 + 2x_2 t + x_3 &= 0, \\ x_2 t^2 + 2x_3 t + x_4 &= 0. \end{aligned} \right\} \qquad 2·63$$

As the ∞^1 tangents of C lie on I and J they are generators of the sextic surface which is common to these primals. But from the fundamental polarity, since six osculating planes meet a general line (2·5), it follows that *six tangents meet a general plane*; the tangent surface is thus the complete intersection of I and J.

The complete figure of C thus has *characteristic numbers* (4, 6, 6, 4), i.e. four points of C are incident with a general prime, six tangents of C are incident with a general plane, etc. The projection of C from a point of itself is the normal curve in [3] whose tangents and osculating planes are projections of those of C, while dually the osculating primes, osculating planes and tangents of C meet a fixed osculating prime in the osculating planes, tangents and points of a normal curve in the fixed prime: this curve is called the *osculant* of C in that prime.

Invariants and covariants

3·1. It will have been observed that the most important loci discovered in connection with the curve, the primals I and J, are given by the vanishing of the two invariants i, j of the quartic form, $x_0t^4 + 4x_1t^3 + 6x_2t^2 + 4x_3t + x_4$, while the vanishing of this form itself gives the relation between a point (x) and a prime (t) which passes through it (2·42). This is an illustration of an essential liaison between the algebra of invariants and the geometry of the rational curve which must now be discussed.

A collineation in [4] transforms a quartic curve C into another quartic curve $\bar{C}$; and any range P_1, P_2, ... on C is transformed into a projective range $\bar{P}_1$, $\bar{P}_2$, ... on $\bar{C}$. If the parameter of P on C is t and of $\bar{P}$ on $\bar{C}$ is $\bar{t}$, then $\bar{t} = (at+b)/(ct+d)$. But, conversely, if C, $\bar{C}$ are quartic curves and P_1, P_2, ... and $\bar{P}_1$, $\bar{P}_2$, ... are given projective ranges on them, then there is a collineation in [4] which transforms C into $\bar{C}$ and the range (P) into the range $(\bar{P})$: for, suppose the curve C is given by $x_i = (-t)^i$ and $\bar{C}$ by $x_i = \Sigma a_{ik}\bar{t}^k$, and the given projectivity is $\bar{t} = (at+b)/(ct+d)$, then by substituting for $\bar{t}$ in terms of t in the equations of $\bar{C}$ these are obtained as $x_i = \Sigma b_{ik}t^k$; and the collineation $\bar{x}_i = \Sigma(-1)^k b_{ik}x_k$ transforms C into $\bar{C}$ and (P) into $(\bar{P})$.

If C and $\bar{C}$ are the same curve C, it appears that a collineation transforming C into itself induces a projectivity on the curve; and that given any projectivity on C there is a unique collineation of C into itself which induces this projectivity. Hence there are ∞^3 collineations of C into itself corresponding precisely to the ∞^3 possible projectivities on the curve, or alternatively to the ∞^3 ways of choosing the osculating system of reference; their formulae are the same as those required in transforming from one osculating system of reference to another osculating system. If in the allied transformations acting on (x) and t, $x \to \bar{x}$ and $t \to \bar{t}$, then the quartic equation

$$x_0t^4 + 4x_1t^3 + 6x_2t^2 + 4x_3t + x_4 = 0,$$

expressing as it does that the point (x) is incident with the prime (t), is transformed into

$$\bar{x}_0\bar{t}^4 + 4\bar{x}_1\bar{t}^3 + 6\bar{x}_2\bar{t}^2 + 4\bar{x}_3\bar{t} + \bar{x}_4 = 0,$$

and the collineation may be written down as the transformation of the coefficients of this equation when the projectivity $t \to \bar{t}$ is given. Hence the vanishing of invariants of the quartic form will give equations of loci which are unaltered under collineations of the curve into itself. But loci which are projectively related to C are precisely those which are unaltered under such collineations, and thus all such loci can be expressed in terms of invariants (and covariants) of this and analogous quartic forms.

The point (x) will be said to *represent* the quartic equation with reference to C, our notation having been chosen so that this equation appears in its standard form. A point t_1 of C itself represents a quartic equation with four roots equal, $(x-t_1)^4 = 0$; a point of a tangent of C represents one with three roots equal, and a point of an osculating plane of C one with two roots equal.

If $F(x) \equiv F(x_0, x_1, x_2, x_3, x_4)$ is an invariant of $(xt)^4$ of degree n, then $F(x) = 0$ is the equation of a primal of order n which can be constructed projectively from the curve. Examples are $i = 0$ and $j = 0$, which are the equations of the primals I and J; these primals are the loci of points whose polars meet the curve in equi-anharmonic and harmonic sets of points respectively. Again, $i^3 - 27j^2 = 0$, being the condition that the quartic equation should have two roots equal and therefore be represented by a point of an osculating plane of C, is the equation of the locus of such planes.

If $F(x, y) \equiv F(x_0, ..., x_4; y_0, ..., y_4)$ is an invariant of degree (m, n) of two quartic forms, the equation $F(x, y) = 0$ introduces between the points (x) and (y) an (m, n) correspondence which is projectively related to C. Corresponding to a particular value of (y), the locus of (x) is a primal of order m related to C and (y); in a collineation of C into itself, $(y) \to (\bar{y})$ and the new locus is projectively related to C and $(\bar{y})$ in the same way as the old locus was related to C and (y). Similarly, if $F(x; t)$ is a covariant of $(xt)^4$, then $F(x; t) = 0$ is a primal related to a point t of C, or more generally we may have $F(x; t_1, t_2, ...) = 0$ related to several points of C. A special case is when this last form is a polarised form of $F(x, t)$; then we replace the symmetric functions of $t_1, t_2, ...,$ by the coefficients of the form having these for roots, $(pt)^m = 0$, and the resulting equation $\Phi(x; p) = 0$, expressing the apolarity of $(pt)^m$ and $F(x, t)$, is a primal projectively related to the set of m points on the curve.

In this way all the formulae of the last section may be interpreted in invariant terms. The trisecant planes of C, for instance, which pass through a point (x) are given by the cubics apolar to the quartic $(xt)^4$. As another typical example consider the hessian of the quartic form: the equation

$$(ht)^4 \equiv (x_0 x_2 - x_1^2) t^4 + \dots + (x_2 x_4 - x_3^2) = 0 \qquad 3{\cdot}11$$

gives a quadric related to a point t of C; to find this relation it is only necessary to examine the case when $t = \infty$ for which the equation is $x_0 x_2 - x_1^2 = 0$; this represents a quadric cone passing through C and having the tangent at $t = \infty$ as vertex. Hence from the covariance of the hessian it follows that 3·11 gives a quadric cone through C with a line vertex which is the tangent at the point (t); it is the cone projecting C from a tangent into a conic. If, on the other hand, (x) is regarded as fixed, the equation 3·11 determines the four points (t), whereat the tangent lies in a plane through (x) which meets C again.

Further, the equation obtained by polarising the hessian, viz.

$$(x_0 x_2 - x_1^2) t_1 t_2 t_3 t_4 + \tfrac{1}{2} (x_0 x_3 - x_1 x_2) \Sigma t_1 t_2 t_3 + \dots + (x_2 x_4 - x_3^2) = 0,$$
$$3{\cdot}12$$

represents a quadric passing through C; and, since for $t_1 = \infty$ it contains $x_0 = x_1 = x_2 = 0$, this quadric contains the four tangents at t_1, t_2, t_3, t_4. If $(pt)^4 = 0$ gives the four points of contact of these tangents, the equation may be written

$$(x_0 x_2 - x_1^2) p_4 - 2 (x_0 x_3 - x_1 x_2) p_3 + \dots + (x_2 x_4 - x_3^2) p_0 = 0.$$
$$3{\cdot}13$$

There are ∞^{14-9} quadrics through C and therefore ∞^1 of them containing four prescribed tangents; this pencil is given by the linear combination of the above quadric with the quadric I which contains all the tangents.

Finally, suppose the covariant relation 3·11 holds for every value of t, then 3·13 gives a linear ∞^4 system of quadrics which has C as basis of the system. Any point common to the quadrics is on C, or, in algebraic terms, a quartic whose hessian vanishes identically is a complete fourth power. The linear ∞^5 system obtained by combining the above with I is equivalent to the linear combination of the quadrics given by

$$\begin{Vmatrix} x_0 & x_1 & x_2 & x_3 \\ x_1 & x_2 & x_3 & x_4 \end{Vmatrix} = 0.$$

This last example shews that not every locus projectively related to the curve can be expressed by an invariant in the strict sense of the term: it may be that it is the basis of a system of loci given by the identical vanishing of a covariant.

A quartic equation is represented with reference to C by a point in accordance with 2·42, but on account of the fundamental polarity it may equally well be represented by the polar prime (ξ) of the point (x) according to the formula

$$(\xi t)^4 \equiv \xi_0 - \xi_1 t + \xi_2 t^2 - \xi_3 t^3 + \xi_4 t^4 = 0,$$

which unites the prime (ξ) with an incident point of C. Then (x) and (ξ) represent the same equation if they are pole and polar. Moreover, if (x) and (ξ) are incident,

$$(\xi x)^4 \equiv \xi_0 x_0 + \xi_1 x_1 + \xi_2 x_2 + \xi_3 x_3 + \xi_4 x_4 = 0,$$

and the quartics represented by (x) and (ξ) are apolar.

Invariants, when expressed in ξ, η, ..., yield the equations of envelopes which are the reciprocal polars of the former loci.

Symbolic notation

3·2. We shall have occasion to use the symbolic notation for invariants* partly as a shorthand and partly as a suggestive method of procedure. Concerning this notation we merely make two remarks on the underlying principles.

(i) By means of the polarising process (of Aronhold) on the coefficients, any invariant can be regarded as linear in the coefficients of each of a set of forms, in number equal to the total degree of the invariant. Then in order to express the original invariant, the requisite number of these forms is said to be the same. This process is the equivalent of replacing the equation of, say, a cubic primal by the trilinearity between the triads of points conjugate to it.

---

     * Grace and Young, *Algebra of Invariants*.

(ii) The symbol x is written for the point (x) *as if* (x) were a point of C, i.e.

$$x_0 : x_1 : x_2 : x_3 : x_4 = 1 : -x : x^2 : -x^3 : x^4,$$

and ξ is written* as a symbol for the prime (ξ) as if (ξ) were a prime of C, i.e.

$$\xi_0 : \xi_1 : \xi_2 : \xi_3 : \xi_4 = \xi^4 : 4\xi^3 : 6\xi^2 : 4\xi : 1.$$

There will be no ambiguity in the resulting symbolic expressions provided they are linear in the coordinates of each point (or prime) introduced and this is effected as above.

The advantage of the symbols lies in the fact that when x and $\bar{x}$ are symbols, $\bar{x} = (ax+b)/(cx+d)$ gives the collineation $x \to \bar{x}$ of the curve C into itself which is allied to the projectivity $t \to \bar{t}$ given by $\bar{t} = (at+b)/(ct+d)$. Hence every interpretable product of the differences of symbols such as $(xy)^4$ [meaning $(x-y)^4$], $(xt)^4$, $(xy)^2 (xt)^2 (yt)^2$, etc., is an invariant (or covariant) of quartic forms, where by "interpretable" is meant that every symbol occurs in the expression four times, and conversely, every invariant (or covariant) can be written in this way. When one form is represented equivalently by different symbols we shall use $x, x', x'', \ldots$ as the equivalent symbols: e.g.

$$(xx')^4 \equiv 2\,[x_0 x_4 - 4x_1 x_3 + 3x_2^2].$$

The notation may be illustrated by writing symbolically the formulae of preceding sections 2·4–3·1.

Prime containing t_1, t_2, t_3, t_4:

$$(xt_1)\,(xt_2)\,(xt_3)\,(xt_4) = 0, \quad \text{or} \quad (x\lambda)^4 = 0, \qquad\qquad 2\cdot41$$

where $t_1, \ldots, t_4$ are the roots of

$$(\lambda t)^4 \equiv \lambda_0 - \lambda_1 t + \lambda_2 t^2 - \lambda_3 t^3 + \lambda_4 t^4 = 0.$$

Trisecant plane:

$$(xt_1)\,(xt_2)\,(xt_3)\,(xT) = 0, \quad \text{or} \quad (x\lambda)^3\,(xT) = 0.\dagger \qquad 2\cdot51$$

Chord:	$(xt_1)\,(xt_2)\,(xT)^2 = 0, \quad \text{or} \quad (x\lambda)^2\,(xT)^2 = 0.$	2·61
Osculating prime:	$(xt)^4 = 0.$	2·42
Osculating plane:	$(xt)^3\,(xT) = 0.$	2·53
Tangent:	$(xt)^2\,(xT)^2 = 0.$	2·63
Quadric I:	$(xx')^4 = 0, \quad \text{or} \quad (\xi\xi')^4 = 0.$	2·43

---

* And Greek letters generally, as opposed to Roman.

† T will be used throughout to indicate that the equality holds for all values of T.

Chordal J: $\qquad\qquad (xx')^2 (x'x'')^2 (x''x)^2 = 0.$ $\qquad\qquad$ 2·62

Hessian of $(xt)^4$: $\qquad (ht)^4 \equiv (xx')^2 (xt)^2 (x't)^2 = 0.$ $\qquad$ 3·11

Polarised hessian: $\qquad (ht_1)(ht_2)(ht_3)(ht_4) = 0,$ $\qquad\qquad$ 3·12

or $\qquad\qquad\qquad (hp)^4 \equiv (xx')^2 (xp)^2 (x'p)^2 = 0.$ $\qquad$ 3·13

Cone of trisecant planes:

$$(xx')^2 (yy')^2 (xy)(xy')(x'y)(x'y') = 0. \qquad\qquad 2·52$$

3·3. The quadrics whose equations can be constructed in this manner may be presented in the scheme:

$$(xa)^4 (x'a)^4 = 0 \qquad \ldots(a), \qquad (\xi\alpha)^4 (\xi'\alpha)^4 = 0 \qquad\qquad \ldots(\alpha),$$

$$(xx')^2 (xb)^2 (x'b)^2 = 0 \quad \ldots(b), \qquad (\xi\xi')^2 (\xi\beta)^2 (\xi'\beta)^2 = 0 \quad \ldots(\beta),$$

$$(xx')^4 = 0 \qquad\qquad \ldots(c), \qquad (\xi\xi')^4 = 0 \qquad\qquad\qquad \ldots(\gamma),$$

and may be interpreted as follows:

3·3,1. Quadrics (c) and (γ) are the same quadric I containing as a locus all the tangents of C and as an envelope all the osculating planes of C. The quadric (b) passes through the curve C and contains the tangents at $(bt)^4 = 0$, while the quadric (a) meets C at the points $(at)^8 = 0$.

2. The quadric (a) is outpolar to (α) if and only if $(a\alpha)^8 = 0$; similarly (b) is outpolar to (β) if $(b\beta)^4 = 0$.

3. When a and b are variable, the linear combination of the systems (a) and (b) with the quadric (c) gives the complete system of quadrics in [4].

4. The linear combination of (b) and (c), (c) and (a), (a) and (b) gives respectively the complete system outpolar to the system (α), (β), (γ). Thus any quadric (a) is outpolar to C and any quadric (α) is inpolar. Through eight points arbitrarily given on C there is one quadric outpolar to C.

5. If P, P' are a pair of points conjugate with respect to a quadric (a), the figure formed by the osculating primes to C from P and P' has all its opposite vertices conjugate with respect to (a); the osculating primes are given by $(\rho t)^8 = 0$, where $(\rho a)^8 = 0$.*

6. The osculating primes given by $(\rho t)^5 = 0$, where $(\rho a)^5 (aT)^3 = 0$, form a self-polar simplex for the quadric (a). There are ∞^1 such simplexes and their vertices lie on a quartic curve to which (a) is inpolar and whose developable is the reciprocal of C with respect to (a).

3·4. As an example of the application of the notation to higher forms related to C the following may be considered:

* Meyer, *Apolarität u. rationale Kurven* (1883) 198.

The cubic primal $(xp)^4 (x'p)^4 (x''p)^4 = 0$ meets C where $(pt)^{12} = 0$, and has C for an apolar curve, i.e. every polar quadric of the primal is outpolar to C. The osculating primes through any of three points which form a conjugate triad for the primal osculate C at $(qt)^{12} = 0$, where $(qp)^{12} = 0$. The septimics apolar to $(pt)^{12}$ give a g_7^1 on C; the osculating primes, $X_i = 0$, at a set of this g_7^1 are a polar 7-ad for the primal, i.e.

$$(xp)^4 (x'p)^4 (x''p)^4 \equiv X_1^3 + X_2^3 + X_3^3 + X_4^3 + X_5^3 + X_6^3 + X_7^3.$$

The general cubic primal, $\psi = 0$, which involves thirty-four constants as compared with the above involving thirty-three, cannot be expressed as the sum of seven cubes, but if a prime $Y = 0$ is arbitrarily prescribed, there is one value of λ for which $\psi + \lambda Y^3 = 0$ has an apolar quartic curve and thus $\psi + \lambda Y^3$ can be expressed as the sum of seven cubes in ∞^1 ways.*

Extension to normal curves of higher order

3·5. Arguments precisely analogous to those of all the preceding sections may be used to deal with a normal rational curve of any order n.

The fundamental polarity is quadric when n is even, and null when n is odd, and is determined in either case by $(xy)^n = 0$ or $(\xi\eta)^n = 0$.

When $n = 3, 5, 7, \ldots$ there is respectively one chord, trisecant plane, quadrisecant solid, ... through a general point, corresponding to the fact that $(xt)^n$ can be expressed uniquely as the sum of nth powers. When $n = 4, 6, \ldots$ there is through a point a quadric cone of trisecant planes, cubic cone of quadrisecant solids, ..., forming a g_3^1, g_4^1, ... on the curve.

The chords of the curve $(n > 3)$ form a three-dimensional locus of order $\frac{1}{2}(n-1)(n-2)$, the trisecant planes $(n > 5)$ form a five-dimensional locus of order $\frac{1}{6}(n-2)(n-3)(n-4)$, etc.

Similarly the result of the next section (4·1) may be immediately generalised; the chords of an involution form a scroll of order $n - 1$ lying on the chordal. But for the most part the results which follow are peculiar to the quartic curve, depending as they do on the simplicity of its invariant system and on the ease with which those invariants can be interpreted.

---

* Cf. Palatini, *Atti Acc. Torino*, **38** (1903) 43.

3·5,1. The characteristic numbers (2·6) of the normal rational curve in $[n]$ are $[n, 2(n-1), 3(n-2), 4(n-3), ..., (n-1)2, n]$.

2. For curves of order n in $[n]$: if $n = 2m+1$, a single $(2m-1)$-fold may be drawn through a general $(m-1)$-fold to meet the curve in $2m$ points; also a single m-fold may be drawn through a general point to meet the curve in $(m+1)$ points—if $n = 2m$, $\infty^1 (2m-2)$-folds may be drawn through a general $(m-2)$-fold to meet the curve in $(2m-1)$ points; also ∞^1 m-folds may be drawn through a general point to meet the curve in $(m+1)$ points.

Quadratic involutions on C: director lines and planes

4·1. A quadratic involution (g_2^1) on C is induced by a collineation of C into itself which must also be involutory. The parameter of the double points may be taken as $t = 0, \infty$, so that the involution is $t \to -t$, and the collineation $x \to -x$, that is,

$$x_0, x_1, x_2, x_3, x_4 \to x_0, -x_1, x_2, -x_3, x_4.$$

This is a harmonic inversion with $x_0 = x_2 = x_4 = 0$ as director line and $x_1 = x_3 = 0$ as director plane. Thus the director line is determined as the join of the points in which the tangents at the double points of the g_2^1 meet the axis plane common to the osculating primes at these points. The director plane is the polar plane of the director line; it passes through the double points and through the point of intersection of the osculating planes at these points.

Any point (x) of a chord of the g_2^1 is

$$[\lambda+\mu, -(\lambda-\mu)t, (\lambda+\mu)t^2, -(\lambda-\mu)t^3, (\lambda+\mu)t^4],$$

so that the locus of (x) is the cubic surface common to the quadric cones

$$\begin{Vmatrix} x_0 & x_1 & x_2 \\ x_2 & x_3 & x_4 \end{Vmatrix} = 0. \qquad\qquad 4\cdot11$$

The chords of the g_2^1 trace out projective ranges upon the director line and the conic $x_0, x_2, x_4 = 1, t^2, t^4$, which passes through the double points and lies in the director plane. Thus *from the director line C is projected doubly into the conic in the director plane.* The prime determined by the director line and any tangent of this conic contains two consecutive chords of the g_2^1; it therefore touches the cubic scroll 4·11 along a chord of C and is bitangent to C at the ends of the chord. *The bitangent primes of the involution envelope the quadric cone*

$$x_0 x_4 - x_2^2 = 0, \qquad\qquad 4{\cdot}12$$

which has the director line as vertex; the chords of the involution generate the normal cubic scroll 4·11 *which is touched along its generators by the bitangent primes, and has the director line as directrix.* As any prime through one of the chords meets the scroll again in a conic passing through two points of C, the involution is cut out by primes through the plane of any conic so determined.

4·1,1. Any quadric through the cubic surface is a cone with its vertex on the surface.

2. If the double points of the g_2^1 are given by $(pt)^2 = 0$, the cubic surface is the intersection of the three quadric cones

$$\begin{Vmatrix} x_0 & x_1 & x_2 & p_0 \\ x_1 & x_2 & x_3 & p_1 \\ x_2 & x_3 & x_4 & p_2 \end{Vmatrix} = 0, \quad \text{i.e.} \quad (xx')^2 \, (xp) \, (x'p) \, (xT) \, (x'T) = 0.$$

3. The quadric whose equation is $(xx')^2 \, (xp) \, (x'p) \, (xq) \, (x'q) = 0$ contains the chords of the two involutions whose double points are respectively $(pt)^2 = 0$ and $(qt)^2 = 0$. This quadric is a cone with its vertex where the common chord of the two involutions meets the prime determined by the four double points.

If $p = q$, the quadric is the line-cone of bitangent primes of the involution; hence the director line is given by the equations

$$p_1^2 x_0 - 2 p_1 p_0 x_1 + p_0^2 x_2 = 0,$$
$$p_1 p_2 x_1 - (p_1^2 + p_0 p_2) \, x_2 + p_0 p_1 x_3 = 0,$$
$$p_2^2 x_2 - 2 p_2 p_1 x_3 + p_1^2 x_4 = 0.$$

The g-lines

4·2. We now discuss the collineations of C into itself which leave a given general point (x) fixed. Under such collineations, however, all the points of a line remain fixed: for, the hessian of $(xt)^4$ is the quartic covariant $(ht)^4$, in general distinct from $(xt)^4$, so that the point (h) representing $(ht)^4 = 0$ and hence all the points $(x + \lambda h)$ of a line remain fixed. We call it the *g-line* of (x); from its construction it is the g-line of any point of itself. Further, the polar primes of points of this line cut out a *syzygetic* involution on C. The characteristic property of the syzygetic pencil of quartic forms $(xt)^4 + \lambda (ht)^4$ is that it contains three members which are complete squares. Thus three points, K_i, of a g-line are points of intersection of pairs of osculating planes.

The collineations of C into itself which leave (x) fixed must leave the cross-ratio of the roots $t_1, \ldots, t_4$ of $(xt)^4 = 0$ unaltered, and thus induce three quadratic involutions which interchange the roots in pairs: $\mathbf{I}_1 = (t_2 t_3)(t_1 t_4)$, $\mathbf{I}_2 = (t_3 t_1)(t_2 t_4)$, $\mathbf{I}_3 = (t_1 t_2)(t_3 t_4)$. These satisfy $\mathbf{I}_2 \mathbf{I}_3 = \mathbf{I}_1, \mathbf{I}_3 \mathbf{I}_1 = \mathbf{I}_2, \mathbf{I}_1 \mathbf{I}_2 = \mathbf{I}_3$; with the identical transformation they form a group usually termed the *axial group*. The allied harmonic inversions similarly present an axial group which is thus the group of all collineations of C into itself which leave a given general point, and consequently its g-line, fixed. The g-line is then common to the three director planes ϖ_i of $\mathbf{I}_i$.

The illustration of these involutions on a conic is well known as arising from a triangle self-conjugate with respect to the conic; we proceed to exhibit them on C.

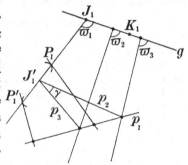

The pairs of double points P_i, $P_i'(i = 1, 2, 3)$ of the involutions $\mathbf{I}_i$ are mutually harmonic. Thus the director plane ϖ_1 contains P_1, P_1' and meets the lines $P_2 P_2'$, $P_3 P_3'$; it therefore passes through the unique transversal of the chords $P_i P_i'$. The director planes ϖ_i thus meet in a line, g, while the director lines p_i form a triangle in the polar plane, γ; the bitangent primes at P_i, P_i' pass through γ, and dually, the osculating planes at P_i, P_i' meet at K_i on g. The chord $P_i P_i'$ passes through the vertex J_i' of the triangle in γ and meets g at J_i.

Given any general point, we can find an osculating system of reference (by taking $P_1, P_1', P_2 = \infty, 0, 1$) such that the points of C on the polar prime of the given point are $\pm t_1$, $\pm t_1^{-1}$, and are thus given by $t^2 + t^{-2} = t_1^2 + t_1^{-2}$. The point then represents a quartic equation in *canonical form*, $t^4 + 6mt^2 + 1 = 0$, and has coordinates $(1, 0, m, 0, 1)$. The hessian equation is of the same form, being
$$t^4 + 6m't^2 + 1 = 0, \quad m' = (1 - 3m^2)(6m)^{-1},$$

and so all the points of the line g through the given point are represented by varying λ in $t^4 + 6\lambda t + 1 = 0$. As the hessian is quadratic in m there are two points (x) which have the same point (h); these points (x) are conjugate in regard to the quadric I.

4·2,1. By proper choice of the system of reference the involutions are obtained as $t \to -t$, $t \to t^{-1}$, $t \to -t^{-1}$ and the allied harmonic inversions are then $x \to x$, $x \to x^{-1}$, $x \to -x^{-1}$; the equations of p_i, ϖ_i are obtained accordingly, while γ, g are found to be

$$\gamma: \qquad x_0 + x_4 = x_2 = 0,$$
$$g: \qquad x_0 - x_4 = x_1 = x_3 = 0.$$

Any point of g has coordinates $(1, 0, \lambda, 0, 1)$, and the points J_i, K_i are given by $\lambda = 0$, 1, -1 or $\lambda(\lambda^2 - 1) = 0$, and $\lambda = \infty$, $-\frac{1}{3}$, $\frac{1}{3}$ or $9\lambda^2 - 1 = 0$. The g-line meets the quadric I in I_1, I_2 given by $1 + 3\lambda^2 = 0$. A simple linear transformation results in J_i being given by $\mu^3 + \nu^3 = 0$, K_i by $\mu^3 - \nu^3 = 0$, and $I_1 I_2$ by $\mu\nu = 0$, shewing that J_i, K_i are two triads with the same hessian points I_1, I_2.

2. As to the number of osculating systems of reference for which a given point has the form $1, 0, m, 0, 1$, there are twenty-four ways of selecting an ordered pair $P_i P_i'$ as ∞, 0 and one further point P_j as unit point; there are thus twenty-four possible reference systems. In sets of four of these the change from one to another is given by the formulae of the axial group above; P_i and P_i' are interchanged, or the point P_j' is taken instead of P_j as unit point. The given point has the same coordinates with respect to each of the set of four. On the other hand on changing from one pair $P_i P_i'$ to another the coordinates of the point are obtained with $24/4 = 6$ different values of m. Changing the unit point from P_j to P_k implies a change of sign in m, so that the six values are roots of a cubic in m^2; they are in fact given by the prescribing of the absolute invariant of the quartic form $i^3 . j^{-2} = (1 + 3m^2) \, m^{-2} \, (1 - m^2)^{-2}$.

3. The coordinates of the given point can be taken as $(1, 0, m, 0, 1)$ in general; the exception is when the point lies on an osculating plane, in which case they may be taken as $(0, 0, 1, 0, 1)$, or as $(0, 0, 0, 1, 1)$ if the point is on a tangent. A point on two non-consecutive osculating planes is not exceptional, being $(1, 0, m, 0, 1)$ with $m = \infty$. On the other hand the g-line through the given point (x) is unique provided the hessian point (h) is distinct from (x), and this is always the case except when (x) is on two non-consecutive osculating planes or on C itself.

The chordal J

5·1. The locus of chords of C is a cubic primal J (2·62); as two chords cannot intersect except at a point of C, only one chord passes through a general point of J. Through a point of C there are ∞^1 chords, and, as these do not all lie in a prime, *each point of C is nodal on J.* From the quadratic involutions on C it follows also that *through each point of J pass two director lines and they lie on J.*

A general prime section of J is a cubic surface with nodes at the four points where the prime meets C. A prime bitangent to C touches J at every point of the chord joining the points of contact: for, the points of contact being taken as $t = 0$, ∞, the prime is $x_2 = 0$ and the section $x_0 x_3^2 + x_4 x_1^2 = 0 = x_2$, which is a cubic scroll having the chord $x_1 = x_3 = 0$ as nodal line. The generators of this section are the director lines of the involutions in which 0, ∞ are corresponding points, i.e. all the lines of J which meet the chord, while the simple directrix $x_0 = x_4 = 0$ is the director line of the involution having 0, ∞ as double points. Conversely, *at any point of J not on C the tangent prime of J touches along the chord through the point and is bitangent to C.* More generally, the polar prime with respect to J (J-polar prime) of any point (a) is given by

$$(xa)^2 (aa')^2 (a'x)^2 = 0$$

and meets C where $(aa')^2 (at)^2 (a't)^2 = 0,$

and is thus the polar prime of the hessian point of (a). When (a) is a point of J, the hessian point is the same for all points of the chord through (a), being the point of intersection of the osculating planes at the points of C on the chord. In this case the g-lines through the hessian point form a pencil meeting the chord. Thus we have a construction for g-lines through a point (x) which coincides with its hessian point: the second point, (a), which has the same hessian point as (x) is then any point of the chord.

The J-polar quadric of any point (a) is given by

$$(xx')^2 (xa)^2 (x'a)^2 = 0;$$

it passes through C, in accordance with the fact that C is nodal on J. Conjugate points with respect to this quadric and lying on

C are related as t, t' in $(ta)^2 (t'a)^2 = 0$. Thus the chords of C lying in the J-polar quadric of (a) form a symmetrical two-two correspondence on C; at the double points of this correspondence $(at)^4 = 0$, so that the quadric contains the tangents at the points of C on the polar prime of (a). There are ∞^4 J-polar quadrics; the complete linear system of quadrics through C is given by varying (a) in the equation

$$\lambda (xx')^4 + (xx')^2 (xa)^2 (x'a)^2 = 0. \qquad 5\cdot11$$

5·1,1. If the coefficients in the envelope equation of a conic on a plane are taken as the coordinates of a point (x) in [5], the point-pairs of the plane are represented by the points of an M_4^3

$$\begin{vmatrix} x_0 & x_1 & x_2 \\ x_1 & x_5 & x_3 \\ x_2 & x_3 & x_4 \end{vmatrix} = 0,$$

and the points of the plane by the points of an M_2^4 (a surface of Veronese) which is double on the M_4^3. The sections of these by the prime $x_5 = x_2$ are the curve C given by

$$\begin{Vmatrix} x_0 & x_1 & x_2 & x_3 \\ x_1 & x_2 & x_3 & x_4 \end{Vmatrix} = 0$$

and its chordal J (v. 14·2).

2. There are ∞^4 quadric cones which pass through C; one with a general point as vertex and ∞^1 with a point on J as vertex. Of these cones ∞^2 have line vertices, the lines being precisely the lines of J.

Equations to the cones have been given as 2·52, and in 4·1, 2.

3. The chords of C which belong to the quadric 5·11 meet C at points related by $(ta)^2 (t'a)^2 + \lambda (tt')^2 = 0$. This two-two correspondence forms a g_3^1 on the curve if $12\lambda^2 = i_a$, and in this case the quadric is a cone, the vertex having $(at)^4 = 0$ as its hessian point. The equation 2·52 may be written

$$(xx')^2 (xa)^2 (x'a)^2 = \tfrac{1}{24} (xx')^4 (yy')^4,$$

where $(at)^4$ is the hessian of $(yt)^4$. If $(at)^4 = [(rt)^2]^2$, the cone is a line cone whose vertex is the chord meeting C at $(rt)^2 = 0$, and its equation is

$$(xx')^2 (xa)^2 (x'a)^2 = \tfrac{1}{3} (rr')^2 (xx')^4.$$

On the other hand the two-two correspondence may be composed of two quadratic involutions, and again the quadric is a cone. The condition for this is $4\lambda^3 - i_a\lambda + j_a = 0$ and the equation 5·11 becomes

$$(xx')^2\,(xa)^2\,(x'a)^2 = \tfrac{1}{8}\,(pq)^2\,(xx')^4,$$

where $(at)^4 = (pt)^2\,(qt)^2 = 0$ are the double points of the involutions. This transforms directly into that of 4·1, 2.

4. There are three points from which four given tangents of C project into a skew quadrilateral.

Quartic surface K

6·1. The most notable geometrical form introduced as an invariant in connection with C is the surface which is the locus of the intersection of osculating planes of C;* we denote it by K. It is the reciprocal of J in the fundamental polarity. That it is a surface corresponds to the fact that there are ∞^2 tangent primes of J, and its properties as an envelope are most readily deduced by examining J. Thus at every point of K there is a tangent plane which is an axis plane of C, and there are ∞^3 tangent primes consisting of the pencils through those tangent planes. The surface passes through C,† and the osculating planes of C are tangent planes. Again, the system of chords on J are striated by taking the chords of quadratic involutions on C; the envelope of the bitangent prime of each involution is a quadric line-cone (4·12). So dually the intersection of osculating planes at corresponding points of the involution traces out a conic of K, and K may be regarded as constructed of these conics.

But to study K as a locus it is more direct to observe that a point of K represents a squared quadratic equation

$$P^2 \equiv (xt^2 + 2yt + z)^2 = 0,$$

while the single quadratic

$$P \equiv (xt^2 + 2yt + z) = 0$$

is represented by a point of a plane π with reference to a conic C_0 whose equations are

$$x, y, z = 1, -t, t^2; \quad \xi, \eta, \zeta = t^2, 2t, 1.$$

Thus the surface K is rational, being mapped uniquely, point for

---

* Non-consecutive osculating planes; the tangents of C do not lie on the surface.
† It meets I and J in C only, the former in $2C$ and the latter in $3C$.

point, on the plane π. The coordinates of any point of K are expressed parametrically as

$$x^2, \quad xy, \quad \tfrac{1}{3}(xz+2y^2), \quad yz, \quad z^2.$$

If five points represented by $P_i^2 = 0$ lie in a prime, there is a relation of the form

$$\lambda_1 P_1^2 + \lambda_2 P_2^2 + \lambda_3 P_3^2 + \lambda_4 P_4^2 + \lambda_5 P_5^2 = 0,$$

which implies in π that the envelope equation of the conic C_0 is of this form, and that therefore the five corresponding points, P_i, lie on a conic outpolar to C_0. Thus to a prime section of K corresponds a conic outpolar to C_0.

To a general plane section of K corresponds the set of intersections of two such conics; thus the order of K is four. Between the four coplanar points of K there is a relation

$$\lambda_1 P_1^2 + \lambda_2 P_2^2 + \lambda_3 P_3^2 + \lambda_4 P_4^2 = 0,$$

corresponding to the fact that the four points on π constitute a polar tetrad for C_0.*

Similarly, if three points of K are on a line there is a relation

$$\lambda_1 P_1^2 + \lambda_2 P_2^2 + \lambda_3 P_3^2 = 0,$$

implying that the corresponding points on the plane π form a self-polar triad for C_0. There are thus ∞^3 trisecant lines of K; they are the g-lines of $4\cdot 2$, for there is a unique g-line through any point not on K and this is trisecant at the points K_i.

There is no line on K, since any two distinct points of K determine either one further point of K in line with them, or none. Indeed, there is no curve of odd order on K: for, any algebraic curve on K is mapped into a curve of order n which meets a conic at $2n$ points, so that the curve on K is of order $2n$.

The points of C correspond to the points of C_0, the quadratic P being in this case a complete square. A line l of points in π, $P_1 + \lambda P_2 = 0$, determines an involution $\mathbf{I}_0$ on C_0: the corresponding range of points on K, $(P_1 + \lambda P_2)^2 = 0$, is quadratic in λ and therefore determines a conic; this lies in the director plane, determined

---

* Algebraically, given three binary quadratic forms there is a unique fourth quadratic form in general, such that there is a linear identity between the squares of these forms.

by the points representing $P_1^2 = 0$, $P_2^2 = 0$, and $P_1 P_2 = 0$, of the corresponding involution **I** on C. Thus the conics of K are as the lines on π: one passes through any two points of K and any two meet in one point. When the points $P_1 = 0$, $P_2 = 0$, which define the line on π, are taken on C_0, then $P_1 P_2 = 0$ represents a point on K; it is seen then that the director plane meets K in a conic and an *isolated point* which is the pole with respect to the conic of the chord joining the double points of **I**. To this isolated point corresponds in π the pole of the line l with respect to C_0; this pole forms with pairs of conjugate points on l self-polar triads for C_0, and thus the trisecants of K through the isolated point form the pencil of g-lines in the director plane.* The isolated point is on the conic only when $P_1 \equiv P_2$. The director plane is then the osculating plane, and the conic of K on it touches C and corresponds in π to a tangent of C_0. The pencil of lines in the osculating plane and through its point of contact are g-lines which are tangent trisecants of K; in particular the tangent of C is a g-line which has three coincident intersections with K and we may say that C *is asymptotic on* K: this property is characteristic for C on K.

Equations of K

6·2. The points (x) of K are those points which coincide with their hessian points: thus immediately, the locus equations of K are

$$\left\| \begin{matrix} x_0 & x_1 & x_2 & x_3 & x_4 \\ h_0 & h_1 & h_2 & h_3 & h_4 \end{matrix} \right\| = 0, \qquad 6·21$$

which give cubic primals through K. Equivalently, the jacobian of $(xt)^4$ and $(ht)^4$ vanishes identically. This jacobian, the so-called *sextic covariant* of $(xt)^4$, we write as $(gt)^6$. Then K is the base of a linear ∞^6 system of cubic primals $(ga)^6 = 0$, where $(at)^6$ is an arbitrary sextic. It is at once verified that the determinants in 6·21, save for numerical factors, are the coefficients of $(gt)^6$, three pairs of them being equal. Further, it can be verified that this linear

---

* Cf. 5·1: the isolated point is the hessian point of any point on the chord.

system is complete,* as indeed may be seen from the fact that no quadric contains K and there is no further invariant of degree three.

We proceed to examine some particular members of this system of cubic primals:

(i) The covariant locus $(gt)^6 = 0$ is related to a point t of C. Its leading term is $g_0 \equiv x_0^2 x_3 - 3x_0 x_1 x_2 + 2x_1^3$, and when $t = \infty$ the locus has for its equation $g_0 = 0$. This is a cubic cone, with a point vertex at $t = \infty$, passing through C. Its prime section, say by $x_4 = 0$, is a cubic scroll of the Cayley type with coincident directrices $x_0 = x_1 = 0$; and the twisted cubic, projection of C, is asymptotic on it. Such then is the character of the locus $(gt)^6 = 0$, the vertex being a general point t.

(ii) This cone may be seen from another point of view. The vanishing of the jacobian of any two forms $(xt)^4$ and $(yt)^4$ gives the double points of the quartic involution

$$(xt)^4 + \lambda (yt)^4 = 0,$$

and thus the points of contact of the osculating planes which meet the corresponding line $(x + \lambda y)$. In the case of the syzygetic involution the jacobian $(gt)^6$ factorises into three mutually harmonic quadratics giving the three pairs $P_i P_i'$ on C, and the corresponding line is a g-line. Thus if we fix (x) in the equation $(gt)^6 = 0$, the values of t are those of the points $P_i P_i'$ on the director planes through (x), while if we fix t the point (x) lies, as we have said, on a cubic cone with vertex t. Thus the generating planes of this cone are the director planes which pass through (t); as each of these meets K in a conic, $(gt)^6 = 0$, *regarded as an equation in x, is that of the cubic cone projecting K from a point t of C.*

(iii) This can be generalised by considering the locus

$$(gt_1)^3 (gt_2)^3 = 0. \qquad\qquad 6\cdot 22$$

When $t_1 = \infty$ and $t_2 = 0$, this reduces to the vanishing of the middle term of $(gt)^6$, i.e.

$$2g_3 \equiv x_1^2 x_4 - x_0 x_3^2 = 0,$$

---

* There are ∞^{27} curves of order twelve on K, this being the number of curves of order six on π. If a cubic primal passes through twenty-eight points of K it must contain it; there are therefore $\infty^{34-28} = \infty^6$ cubic primals containing K.

which is the equation of the cubic cone generated by the director planes through $(0, 0, 1, 0, 0)$, this cone being the polar of the section of J by its tangent prime. *Thus* 6·22, *or* $(ga)^2 (ga')^2 (ga'')^2 = 0$, *as an equation in* (x), *is the cubic cone projecting* K *from that point of itself which represents the quadratic* $(at)^2 = 0$. But this cone may also be regarded as generated by the ∞^2 g-lines which pass through the points of a conic of K, for these are in pencils on the director planes through the isolated point, $(at)^2 = 0$, of the plane of the conic. In fact the g-lines which meet a director plane or lie in it generate the cubic cone 6·22.

Returning now to the general case we observe from the equation

$$(ga)^6 \equiv (xx')^2 (xx'') (xa) (x'a)^2 (x''a)^3 = 0$$

that the tangent prime to this primal at the point t_0 of C is given by

$$(xt_0)^3 (xa) (at_0)^5 = 0$$

and meets C again at t, where $(at) (at_0)^5 = 0$. The tangent prime thus meets C three times at its point of contact (contains the osculating plane there), while its orientation is determined by the linear polar of the point with respect to $(at)^6 = 0$. At each of the six points given by $(at)^6 = 0$ the tangent prime is the osculating prime to C; the curve is thus asymptotic on the primal at these points and the tangents thereat to C lie on the primal.

6·2, 1. The tangent prime of $(ga)^6 = 0$ at any point $(ct)^2 = 0$ of K is the particular tangent prime of K thereat which passes through the point representing $(ac)^2 (at)^4 = 0$.

2. The polar quadric of t_0 on C with respect to $(ga)^6 = 0$ passes through C and contains the tangents at t_0 and at the three points given by $(at_0)^3 (at)^3 = 0$.

The system of Segre cubic primals

6·3. Of the ∞^6 cubic primals through K one may be drawn to contain an arbitrary plane α: for, if it is drawn through six general points of α it passes also through four further points of α (on K) and thus contains α. Only one primal can be drawn through α, except in the case when α is a director plane and there are then ∞^4 primals. If we leave aside the exceptional case which is that dealt with as 6·22, a g-line which meets α, since it contains four points of the primal, lies on it. Hence the ∞^2 g-lines which meet α generate the primal. In general any six g-lines determine uniquely a cubic primal containing them and K, and this is the primal through a plane α which meets the six lines. In particular the six tangents to C at $(at)^6 = 0$ determine the primal $(ga)^6 = 0$ and it passes through any plane meeting the six tangents.

The four points of α on K are nodes of the primal, for any line through such a point meets the primal twice there.* A prime through the plane α meets the primal in a quadric through the four points (K, α); g-lines of the primal on this quadric form a regulus and are thus met by one line through each of the four points; so that by taking the pencil of primes through α, we see that *the g-lines which meet a plane α meet also four other planes and generate a cubic primal containing K and the five planes, and having ten nodes at points of K where these planes meet in pairs.* These primals are thus Segre cubic primals: they have the greatest possible number of nodes for a cubic primal which does not degenerate. The five planes are *associated* in the sense that any line meeting four of them meets the fifth, and the lines meeting them are one system of generators of the primal.

The g-lines which meet two arbitrary planes (not on the same cubic primal through K) lie on the residual intersection of two primals which is a quintic scroll, W; and there are five (associated) g-lines which meet three arbitrary planes.

Configuration of nodes on K

6·4. At each node of the cubic primal there is a pencil of g-lines which must meet each of the three planes which do not pass through the node in the points of a line: each of these lines must pass through two nodes. Thus there are ten director planes on the primal and the conics of K in these planes pass each through three nodes. If the original five planes be denoted by 1, 2, ..., 5, and the nodes at their intersections by 12, 13, ..., 45, the ten conics may be denoted by 123, 124, ..., 345, where the conic 123 passes

---

* In the neighbourhood of the point the lines which meet both α and K meet the primal twice, and do not lie in one prime since α meets the tangent plane to K in one point.

through 12, 13, 23, and 45 is the isolated point of K in its plane. This configuration of the nodes on K, when mapped on a plane π, is the Desargues configuration of ten points and ten lines, in which three points lie on a line and three lines pass through a point, the figure being self-polar with respect to C_0. Further, the notation suggests the consideration of the Desargues configuration as the plane section of the figure of five general points 1, 2, ..., 5, in [3]. But now by means of the ∞^4 quadrics through these five points, we can map the space [3] so that the quadrics become primes in [4], and find that to every point in [3] corresponds a point of a Segre cubic primal, and conversely—except that to each of the five points themselves corresponds a plane of the primal. The lines joining the points 12, ... become the nodes of the primal and the planes 123, ... the further ten planes. The ∞^2 cubic curves through the five base points correspond to the ∞^2 g-lines of the original figure. To a general plane π corresponds a quartic surface having trisecants corresponding to the triads in which the cubic curves meet π: as these triads are self-polar for a conic C_0 of the plane, it is clear that our representation includes our former mapping of K on π. We also note that on any Segre cubic primal there are ∞^3 quartic surfaces of the same character as K, all passing through the nodes and having ∞^2 trisecant lines in common; the system is residual on the primal to the ∞^5 quintic scrolls W, and each curve (K, W) is of order ten.

6·4,1. A quartic involution (g_4^1) has six double points. If six distinct points are prescribed there are five g_4^1 having these six points as double points; they correspond to the five associated planes which meet six tangents of C. For no position of the six tangents is there an infinite number of planes meeting them.

Should the six prescribed points have one, two or three pairs of coincident points the number of involutions without fixed points reduces to three, two or one, respectively, the remaining involutions having either one or two fixed points. The corresponding Segre cubic primal has one, two or three of its nodes on C.

Should four or five of the prescribed points be coincident, the total number of involutions in each case is less than five: there is a cubic primal with two nodal lines replacing the Segre cubic primal as the locus of g-lines meeting one of the planes. In the cases where the prescribed points coincide in threes, or all six points are coincident, the cubic primal is a cone [cf. 6·2 (iii) and (i)].

2. In general there is no projectivity on C which transforms a sextic $(at)^6 = 0$ on C into itself, so there is no collineation of C into itself which carries the primal $(ga)^6 = 0$ into itself. If however $(at)^6$ is capable of linear transformation into itself, then there is a corresponding collineation of C into itself which implies some permutation of the five associated planes and carries $(ga)^6 = 0$ into itself. For example, if $(at)^6$ is of g-character, it is capable of an octahedral group G_{24} of transformations into itself; in this case one of the five planes is a g-plane and the corresponding collineations of C into itself are those which leave this one plane unaltered and permute the other four in every possible way.*

3. If six chords of C meet it in pairs of points given by the icosahedral duodecimic equation $t \, (t^{10} + 11t^5 - 1) = 0$, there is a group of sixty collineations of the curve into itself permuting the chords. The five planes meeting the six chords are g-planes and the collineations effect the even permutations of these five planes.

Representation of g-lines: the manifold G

6·5. By means of the cubic primals through K, the space of four dimensions can be mapped so that the primals become primes of [6]. To every general point of [4] corresponds a point of [6], but to this point of [6] correspond all the points of a g-line in [4], since all the cubic primals meeting at a point contain the g-line which is determined by that point. The same mapping is achieved more simply by supposing the points of [6] to represent sextic equations with reference to a normal rational sextic curve C^6. Among these sextic equations are those of g-character, that is, sextics which are the product of three mutually harmonic quadratics. As there are ∞^3 such equations, the points of g-character form a three-dimensional manifold which we denote by G; each point of G, as it represents $(gt)^6 = 0$, corresponds to a g-line in [4]. Then $(ga)^6 = 0$ is the section of G by the prime $(at)^6 = 0$ and corresponds to a cubic primal through K, while $(ga)^6 = (gb)^6 = 0$ is a [4]-section of G corresponding to a quintic scroll, W, of g-lines

---

* Klein, *Vorlesungen über das Ikosaeder* (1884), p. 54. In Klein's notation $w^2 = 0$ represents four nodes on the invariant g-plane, and $\chi^2 = 0$ the remaining six nodes. For other possible sextics v. Bolza, *Am. J. of M.* 10 (1888) 50.

meeting two planes, and $(ga)^6 = (gb)^6 = (gc)^6 = 0$ is the set of five points in a [3]-section of G corresponding to the five g-lines which meet three planes. The order of G is therefore five; the fact that the five g-lines are associated corresponds to the fact that any four points of G determine a solid which meets G again at a fifth point, or algebraically that, given four sextics of g-character, there is a unique sextic of this character which is linearly dependent upon the four.

To a point of K with a pencil of g-lines through it *corresponds a line on G*, the intersection of the ∞^4 primes corresponding to the cubic primals containing the director plane. Through every point of G there are three of these lines; the solid containing the triad is tangent to G, and thus any tangent [4] meets G in a triad of lines and a conic. The points of this [4]-section correspond to the g-lines meeting two planes which have a g-line in common, so that *the conic on G corresponds to the regulus of g-lines which lie in a solid*; there are ∞^2 conics through every point of G, and one conic passes through two arbitrary points. The tangent solids along a line of G lie in one prime; the section of G by this tangent prime corresponds to the lines of the cubic cone which projects K from a point of itself.

6·5, 1. A tangent of C^6 is a line of G. When two of the lines of the triad through a point of G coincide, the point is on the tangent surface of C^6, the double line being a tangent; when all three lines coincide, the point is on C^6. A conic plane of G is the unique transversal of four tangents of C^6.

2. A bitangent solid of C^6 meets G in a line, and the osculating primes at each point of contact contain the line. The prime containing the osculating planes at these points is the tangent prime of G along the line.

3. Through a point of G there are ∞^1 tetrasecant solids of C^6 giving on it a syzygetic pencil whose jacobian is represented by the point of G; the three bitangent solids pass each through a line of G through the point.

4. In reference to the curve C^6 the quadric $(xx')^4 (xp)^2 (x'p)^2 = 0$ contains all the tangents to the curve and also the four osculating planes at $(pt)^4 = 0$. Since it is characteristic of a g-form that $(g, g')^4 = 0$, the ∞^4 quadrics obtained on varying p are the quadrics containing G. The quadric $(xx')^6 = 0$ meets G in the tangent surface (V_2^{10}).

5. In the case when $(pt)^4$ has $(gt)^6$ as its sextic covariant, $(g, p)^2 = 0$; hence every quadric $(xx')^4 (xp)^2 (x'p)^2 = 0$ is a cone, with vertex on G, containing the osculating planes of C^6 at four points whose solid passes through the vertex. With any given vertex on G there is a pencil of such cones of which three are line-cones with vertices on G.

6. From a tangent to C^6 the curve projects into a quartic curve in [4] and G into its quadric I. From a conic of G the curve projects into a four-cusped sextic curve in [3].

7. There is one line of G meeting two given lines of G. There are four lines of G meeting both of two conics of G. The conics of G which meet three given lines of G all meet a fourth line of G.

8. In an arbitrary prime there are ten lines of G, corresponding to the ten nodes of the Segre cubic primal in [4]; if these be denoted by 12, 13, ..., 45, the line 12 meets the lines 34, 45, 53 but does not meet the remaining six.

9. The quintic curve which is the [4]-section of G and consequently the scroll W, is elliptic. The general quadric section of G is a surface of order ten whose prime sections are canonical curves of genus six.*

Linear complexes containing C

7·1. In space of four dimensions a line (or plane) is given by ten homogeneous coordinates with five quadratic relations between them.† We have seen that three pairs of the coordinates of a g-line are equal, save for numerical factors (6·2); *the g-lines therefore belong to three linear complexes* and are thus the common lines of a net of such complexes, which is in immediate agreement with the fact that one g-line passes through a general point. This net of complexes or null polarities when written symbolically is given by
$$(xy)^3 (x\alpha) (y\alpha) = 0, \qquad (\alpha)$$
where the coefficients of $(\alpha t)^2$ are the parameters of the net. The equation (α) may be interpreted directly: the left-hand side of (α) is linear in (x) and (y) and changes sign when x and y are interchanged, so that, for any value of the parameters, (α) expresses a null polarity, and since $(xh)^3 (xT) (yT) = 0$‡ the corresponding complex contains any g-line.

* Scorza, *Ann. di Mat.* (3) 15 (1908) 217, where a full discussion of the manifold G may be found.

† These relations, though linearly independent, are equivalent to three conditions as there are ∞^6 lines.

‡ $= (x'x'')^2 (x'x)^2 (x''x) (x''T) (xT) = 0$ by reason of the factor $(x''x)$.

In particular the tangents to C belong to the complexes (α), and as they do not belong to any other linear complex we say that (α) denotes the complexes *containing* C. With regard to any one of these complexes, the null prime of a point t of C,

$$(xt)^3 (x\alpha) (t\alpha) = 0,$$

meets C three times at t and once at t', where $(\alpha t)(\alpha t') = 0$. The chords of C which are lines of the complex are thus the chords of a quadratic involution on C having $(\alpha t)^2 = 0$ for double points. The null prime of either of these points meets C four times at the point itself, and thus while the null plane of any tangent passes through the tangent, in the case of the tangent at a double point it is the osculating plane thereat. Now any linear complex in [4] is singular and its singular point or *centre* lies on every null prime. Thus the centre of (α) must be on each of the osculating planes at $(\alpha t)^2 = 0$ and is their point of intersection. *The centres of the complexes containing C are the points of K*; while the centres of those which also contain a prescribed line lie on a conic of K.

Since the number of disposable constants in a general net of complexes is the same as that in a quartic curve, namely twenty-one, it is to be presumed that the complexes containing C form a general net, and that *any general net is the system of complexes containing a quartic curve*: this we proceed to shew to be the case.

The lines which meet a fixed plane form a linear complex, the coefficients in its equation being the coordinates of the fixed plane. Such a complex is called *special*, and the fixed plane its *nucleus*; its centre is indeterminate. It is to be remarked then that no complex of a general net is special, as this would involve three conditions in the coefficients; likewise no special complex contains C. Again, the centre of a complex $\Sigma a_{ij} p_{ij} = 0$ is given by the equations $\sum\limits_{j=0}^{4} a_{ij} x_j = 0$ $(i = 0, 1, \ldots, 4)$, and if A_{ij} is the adjoint of a_{ij} in the skew symmetric determinant $|\, a_{ij}\,|$, the coordinates of the centre are proportional to $\sqrt{A_{ii}}$ (since $A_{0i} = \sqrt{A_{00} A_{ii}}$), while the conditions that the complex should be special are given by

$$\sqrt{A_{00}} \equiv a_{12} a_{34} + a_{13} a_{42} + a_{14} a_{23} = 0, \text{ etc.}$$

For a net of complexes, then, the locus of the centre is a surface, the coordinates of the centre being five quadratic polynomials in the two parameters of the net. This surface may be represented on a plane by means of the parameters of the net, and to any prime section of it corresponds on the plane a conic outpolar to a fixed conic, which is not degenerate as no member of the general net is special. If the point equation of the fixed conic is taken as $\alpha_0\alpha_2 = \alpha_1^2$, the general system of conics outpolar to the fixed conic is obtained by linear combination of conics given by $c_i = 0$, where c_i are the coefficients in

$$(ct)^4 \equiv (\alpha_0 t^2 + 2\alpha_1 t + \alpha_2)^2.$$

The centre surface is thus projectively equivalent to the surface $x_i = c_i$, which is the K surface of a quartic curve and indeed precisely the locus of the centres of the net of complexes (α). The common lines of the general net are the trisecants of its centre surface, and thus the general net is projectively equivalent to the net (α).

Similarly, the net of plane complexes

$$(\xi\eta)^3 (\xi a)(\eta a) = 0 \qquad\qquad (a)$$

is a general net. The planes common to the net are the polar planes of the g-lines: these we call the g-planes. The central prime of the complex (a) is bitangent to C at the points $(at)^2 = 0$, and the central primes of the net envelope the primal J.

Linear complexes apolar to C

7·2. We now consider the complexes or null polarities given by

$$(xy)(x\beta)^3 (y\beta)^3 = 0, \qquad\qquad (\beta)$$

$$(\xi\eta)(\xi b)^3 (\eta b)^3 = 0. \qquad\qquad (b)$$

The linear combination of (α) and (β) gives all the ∞^9 complexes in [4], since the members are linearly independent. The two complexes (β) and (b) are conjugate when their sole lineo-linear invariant $(b\beta)^6$ vanishes; similarly (α) and (a) are conjugate when $(a\alpha)^2 = 0$. Hence the ∞^6 system (β) is the complete system conjugate to the net (a) containing C; following the nomenclature used in the case of quadrics (3·3) we call (b) *inpolar* to C and (β)

outpolar to C. The system of inpolar (outpolar) complexes is thus a general linear ∞^6 system.

The chords of C which belong to (β) meet C at points related by

$$(t\beta)^3 (t'\beta)^3 = 0,$$

which is a symmetrical three-three correspondence. The tangents at the six points $(t\beta)^6 = 0$ are the six tangents of C which belong to the complex; all the complexes containing these tangents are given by varying (α) in

$$(xy)(x\beta)^3 (y\beta)^3 + (xy)^3 (x\alpha)(y\alpha) = 0.$$

Of these a certain number will be special, their nuclei being the planes which meet the six tangents; there are thus five special members, and the g-lines which belong to (β), being the common lines of the system, are the lines meeting those five planes. Thus the lines common to all complexes containing six given tangents of C are generators of a Segre cubic primal, whose equation is

$$(g\beta)^6 \equiv (xh)(x\beta)^3 (h\beta)^3 = 0,$$

where $(\beta t)^6 = 0$ gives the contacts of the tangents.

It appears that the lines common to a general linear ∞^3 system of complexes are one system of generators of a Segre cubic primal. In particular the g-lines which meet an arbitrary plane are generators of the Segre primal containing the six tangents of C which meet the plane. Suppose we write

$$(pt)^6 \equiv (xy)(xt)^3 (yt)^3 = 0,$$

$$(\pi t)^6 \equiv (\xi\eta)(\xi t)^3 (\eta t)^3 = 0.$$

These equations give the points of contact on C of the osculating planes meeting the line $p \equiv \overline{xy}$ and of the tangents meeting the plane $\pi = \overline{\xi\eta}$ respectively. Then $(p\pi)^6 = 0$, when π is fixed, is the condition that p should belong to an outpolar complex containing the tangents which meet π, and when p is fixed it is the condition that π belongs to an inpolar complex containing the osculating planes which meet p. If the inpolar complex is special its nucleus must be a g-line; hence $(g\pi)^6 = 0$ *is the form of any special inpolar complex, and is the condition that the g-line meets* π. When the condition holds for π it holds also for the other four planes meeting the same six tangents. On the other hand, if this condition holds for four g-lines, so that π belongs to four (special) inpolar complexes, then π is the generating plane of a Segre cubic envelope and *the line associated with four g-lines is a g-line*; in particular the line associated with four tangents of C is a g-line. Moreover, if we begin with three arbitrary planes π_1, π_2, π_3, the g-lines meeting them are five associated lines and the ∞^2 planes π determined by $\pi \equiv \lambda_1 \pi_1 + \lambda_2 \pi_2 + \lambda_3 \pi_3$ are the generating planes of a Segre cubic envelope. Similarly three arbitrary lines initiate the generating system of a Segre cubic locus containing five g-planes; it meets C at twelve points.

The g-lines which meet two arbitrary planes π_1, π_2 belong to ∞^4 complexes and form a scroll W_2^5. The conjugate system of ∞^4 inpolar complexes have in common the planes $\pi \equiv \lambda_1 \pi_1 + \lambda_2 \pi_2$, forming the dual manifold W_3^5. Each line of W_2 meets each plane of W_3. Two planes of W_3 meet in a point of W_2 and two lines of W_2 determine a solid of W_3; thus W_2 is the double surface of W_3, and W_3 is double (tangentially) on W_2.

7·2, 1. The centres of the ∞^3 complexes containing six tangents of C lie on the Segre cubic primal determined by these tangents.

2. Six general lines in [4] determine a Segre cubic primal through them. Any quartic curve touching the six lines, and its K surface, lie in the primal.

3. There are ∞^3 quartic curves having six given lines as g-lines. Given seven lines there is one quartic curve having them as g-lines.

4. The centre of the complex (β) is the point representing $(\beta\beta')^4 (\beta t)^2 (\beta' t)^2 = 0$, and is indeterminate when β is of g-character and the complex is $(g\pi)^6 = 0$.*

5. Given any sextic involution (g_6^1) on C, the osculating planes of a set are met by five lines; the locus of these lines is a quintic scroll. The locus similarly defined by a g_6^2 is a Segre cubic primal.

Representation of the lines of [4]

7·3. The representation given in 6·5 of the g-lines by points of a manifold in [6] may be presented directly by using the co-ordinates of the lines.

---

* Brusotti, *Ann. di Mat.* (3) 9 (1904) 329. Also cf. 6·5, 4.

Let a plane (line) complex of [4] be represented by a point (prime) of [9], the coefficients of the complex being taken as coordinates; two complexes are then conjugate when the corresponding point and prime are incident. Special plane (line) complexes, or equivalently the lines (planes) which are their nuclei, appear as points (primes) of the grassmannian manifold G_6 (Γ_6) which belongs to five linearly independent quadrics, and is of the fifth order (class), since there are five special complexes conjugate to a linear ∞^3 system. The details of the representation are best seen when set down summarily* in parallel columns thus:

In [4]	In [9]
Plane complex } Line complex } conjugate	Point } Prime } incident
Line } Plane } incident	Point of G_6 } Prime of Γ_6 } incident
Lines through a point	Points of a gen-[3] of G_6
One line passes through two points	Two gen-[3] meet in a point, ∞^1 gen-[3] through the point
Planes of a solid	Primes of a gen-[5] of Γ_6
One plane lies in two solids	Two gen-[5] meet in a plane
Lines of a solid	Points on G_6 of a gen-[5] of Γ_6 $\equiv$ Points of a quadric Ω of G_6

This is the ordinary grassmannian representation of the lines of [3] by the points of a quadric Ω in [5], whence

Point of the solid	Gen-plane α of Ω α is the section of a gen-[3]
Plane of the solid	Gen-plane β of Ω β is the section of a gen-[5] Every prime of Γ_6 touches Ω (or contains it)

As the primes of Γ_6 are those belonging to five linearly independent quadric envelopes, it is seen that the quadrics Ω form a linear ∞^4 system on G_6 and the common tangent primes of this system are the primes of Γ_6. Dually the points of G_6 belong to a

---

* For the representation of the lines of [4] v. Todd, *Proc. L.M.S.* (2) 30 (1930) 513.

linear ∞^4 system of quadric cones whose vertices are gen-[3] of G_6.

The introduction of the curve C in [4] now leads to the following related elements of [4] and [9]:

Plane complex containing C	Point of a plane γ, not meeting G_6
Outpolar complex	Prime through γ
Inpolar complex	Point of a [6] dual to γ
g-line	Point of G, a general [6]-section of G_6
Pencil of g-lines	Line of G, section of gen-[3]
g-lines of a solid	Conic of G, section of gen-[5]
Tangent to C	Point of C^6 on G
g-lines on an osculating plane	Points of a line of G tangent to C^6

We have thus returned to our former representation (6·5) and see, as before, that G is the intersection of a linear ∞^4 system of quadric cones whose vertices are on G (6·5,5). If the [6]-section of primes through γ is taken to represent outpolar complexes, the inpolar and outpolar complexes are represented by the points and primes of [6] and are conjugate when the points and primes are incident; polarity with respect to C in [4] becomes polarity with respect to C^6 in [6]. Each inpolar complex (b) is represented by the points $(bt)^6 = 0$, and each outpolar complex (β) by the prime $(\beta t)^6 = 0$; when $(b\beta)^6 = 0$ the point and prime are incident, and the complexes conjugate. A prime section of G, $(g\beta)^6 = 0$, represents one system of generators of a Segre primal, while a [4]-section $(g, \beta + \lambda\beta')^6 = 0$, an elliptic curve of order five, represents a scroll W. The primes of Γ_6 through the [4] represent the conjugate system of W; they meet the [6] in a pencil of primes which cut out the involution $(\beta + \lambda\beta', t)^6 = 0$ on C^6. Through a [3]-section of G, which contains five points of G, pass ∞^2 primes of Γ^6 representing the planes meeting the five g-lines; these primes meet [6] in a system cutting out the involution $(\beta + \lambda\beta' + \mu\beta'', t)^6 = 0$ on C^6 (7·2,5).

7·3, 1. In general two quartic curves in [4] have five g-lines in common; two K surfaces have five common trisecants.

2. If two quartic curves are contained in a pencil of linear complexes, their common g-lines form one system of generators of a Segre primal.

CHAPTER II

RATIONAL QUARTIC CURVE C_1 IN [3]

Quartic curves of the first and second kind

8·1. Any quartic curve in [3] lies on at least one quadric, since a quadric described through nine points of the curve must contain it. A quartic curve which lies on two quadrics and is thus the base of a pencil of quadrics is said to be of the *first kind*. A quartic curve through which only one quadric can be drawn is said to be of the *second kind*.

A quartic curve of the second kind is rational, since the system of all quadrics cuts out on it a g_8^8. A quartic curve of the first kind on a particular quadric Q is met twice by each generator of either system on Q, and is said to be [2, 2] on Q. In consequence, to every line of one regulus on Q there correspond two lines of the other regulus, thereby establishing a two-two correspondence between the lines of the two reguli; hence there are four lines of each regulus which touch the curve, and thus four double points of the g_2^1 cut out by the generators of either regulus. The curve of the first kind is therefore not rational, unless it has a node, in which case two of the double points are absorbed in the node, and planes through that point cut out a g_2^2 on the curve. Thus a quartic curve with a node (or cusp) is rational although always of the first kind. It follows that the quartic curve of the second kind, being rational and without a node, must be [3, 1] on the quadric through it, and its points are in one-one correspondence with the unisecant generators.

8·1,1. Two quartics of different kinds, or two quartics of the first kind, on the same quadric intersect in eight points. Two quartics of the second kind on the same quadric intersect either in ten or in six points.

2. If a quadric and cubic surface intersect in a conic or in a pair of skew lines, their residual intersection is a quartic curve which is of the first kind or of the second kind respectively (*v.* 12·1,4).

Fundamental quartic f

9·1. The rational quartic curve in [3] will be denoted by C_1, being the projection of a rational normal curve C; it is either of the second kind, or nodal, the latter case arising when the vertex of projection is on the chordal J. Its equations are given by equating the coordinates of any point of it to four linearly independent quartic polynomials in t,

$$x_i = (\alpha_i t)^4 \quad (i = 0, ..., 3), \qquad 9·11$$

and then t is the parameter of the point on C_1.

If four points of C_1 are coplanar, there is a relation between their parameters t_i $(i = 1, ..., 4)$ which is a symmetrical algebraic relation, linear in each parameter, viz.

$$(ft_1)(ft_2)(ft_3)(ft_4) \equiv f_0 t_1 t_2 t_3 t_4 + f_1 \Sigma t_1 t_2 t_3 + f_2 \Sigma t_1 t_2 + f_3 \Sigma t_1 + f_4 = 0,$$

since each parameter is determined uniquely when the other three are given. This relation is the polarised form of the quartic equation

$$(ft)^4 \equiv f_0 t^4 + 4 f_1 t^3 + ... + f_4 = 0,$$

which we call the *fundamental quartic* (*f*-quartic) for C_1. If the parameters t_i are the roots of the equation

$$(et)^4 \equiv e_0 t^4 + 4 e_1 t^3 + ... + e_4 = 0,$$

then $(ef)^4 = 0$; or in words, *any set of coplanar points of C_1 is given by a quartic equation apolar to the fundamental quartic equation, and conversely*. The fundamental equation is represented in [4], with reference to C, by the vertex of projection from which C is projected into C_1, the coplanar points of C_1 being the projection of points of C which are in a prime with the vertex. It is determined immediately from the four quartic polynomials defining C_1 as the unique quartic apolar to them, that is, satisfying the relations $(f\alpha_0)^4 = (f\alpha_1)^4 = (f\alpha_2)^4 = (f\alpha_3)^4 = 0$.

The importance of the *f*-quartic lies in the fact that the effect of any collineation on C_1 is to transform it into another quartic curve $\bar{C}_1$ with the same *f*-quartic equation. Conversely, two rational quartic curves with the same fundamental equation are projectively equivalent: if the points with the same value of the parameter are taken to correspond, there is a one-one correspondence between sets of coplanar points and thus between planes, which establishes a collineation transforming the one curve into the other. Then any projective feature of the curve must be expressible in terms of the f-quartic, and as its expression is independent of the reference system for the parameter on C_1 it will appear as an invariant or covariant of the f-quartic.

The collineation carrying C_1 into $\bar{C}_1$ is not unique, for there are certain collineations which transform C_1 into itself. These latter collineations govern the projective geometry on the curve, and are determined by the linear transformations of the parameter which leave the f-quartic unchanged. Only those transformations of the parameter $t \to \bar{t}$ which leave $(ft)^4 = 0$ unaltered are induced on C_1 by collineations in the surrounding space. In fact only those collineations in [4] of C into itself which have the vertex of projection as a fixed point yield collineations in [3] of C_1 into itself. Such collineations have already been discussed in 4·2 and the results which follow may be obtained directly by projection.

In the general case* the linear transformations $t \to \bar{t}$ which leave $(ft)^4 = 0$ unaltered are three involutions I_1, I_2, I_3 which we call the *principal involutions* on C_1. The corresponding collineations must be involutory. Further, each must be a harmonic inversion: for, if C_1 were transformed into itself by central perspective it would lie on a quadric cone and so be a curve of the first kind—a case here excluded.* *Thus the possible collineations of C_1 into itself are three harmonic inversions*; they form with the identical transformation an axial group.

Principal involutions

9·2. The reference points for the parameter on C_1 being chosen so that the principal involutions are $t \to -t$, $t \to t^{-1}$, $t \to -t^{-1}$, the fundamental equation takes the canonical form

$$t^4 + 6mt^2 + 1 = 0,$$

---

* Special cases require further consideration and are dealt with later (15·1–15·6), they occur (1) when two roots of the f-equation are equal; (2) when the invariants i or j of the f-equation vanish, in which case there are further involutions. The case $j = 0$ is that in which C_1 has a node, as it is the projection of C from a point of J.

and the condition for coplanar points becomes

$$t_1 t_2 t_3 t_4 + m \Sigma t_1 t_2 + 1 = 0. \qquad 9\cdot21$$

The double points of the involution $\mathbf{I}_1$ are 0, ∞ and they are coplanar with t and $-t$. Thus all chords of the involution $\mathbf{I}_1$ meet the chord joining the double points of $\mathbf{I}_1$; accordingly this latter chord is one of the axes of the harmonic inversion corresponding to $\mathbf{I}_1$, the other axis being the polar line of this chord with regard to the quadric on which C_1 lies. These two lines will be called respectively a *principal axis* (or *chord*) and a *principal directrix* of C_1. It is at once verified, by using $9\cdot21$, that the three principal axes meet at a point—the *centre* of the curve; and that the three principal directrices lie in a plane—the *principal plane*—which is the polar plane of the centre with regard to the quadric. Also each axis meets two of the directrices and each directrix two axes, so that they form the edges of a tetrahedron—the *principal tetrahedron* for C_1—which is self-conjugate with respect to the quadric. Any set of four points t, $-t$, t^{-1}, $-t^{-1}$ on C_1 are the vertices of a tetrahedron whose opposite edges meet the opposite edges of the principal tetrahedron in pairs; such sets belong to the syzygetic quartic involution given by varying p in

$$t^4 + 6pt^2 + 1 = 0.$$

Regarding the principal plane as the plane at infinity and the principal axes as mutually orthogonal, we obtain the simplest projective model of the curve; the points $-t$, t^{-1}, $-t^{-1}$ are the images of the point t in the three axes, and the curve is symmetrical with respect to each of its axes.*

The three harmonic inversions will in future be referred to as the *symmetries* of the curve; and any curve or surface or algebraic form will be said to be *symmetrical* when it is unaltered, as is the curve itself, under these symmetries. *Any curve or surface projectively related to C_1 is transformed into itself under the axial group*

---

* Some idea of the shape of the curve is given if a wire taken first in the form of an ellipse on a horizontal plane has each quadrant bent alternately up or down out of the plane while the ends of the two principal axes remain fixed and the wire is made to curve continuously at these ends; in such a figure the principal axis of the curve which is not one of those of the ellipse meets the curve at imaginary points.

*arising from the symmetries of C_1.** For example, the quadric upon which C_1 lies is symmetrical, the four points given by

$$t^4 + 6pt^2 + 1 = 0$$

form a symmetrical tetrad, etc.

The principal tetrahedron presents a natural system of reference for C_1 and its projectively allied curves and surfaces, as their equations referred to it take symmetrical forms. Since three faces of the tetrahedron meet C_1 at two pairs of double points of the principal involutions, and the fourth face meets it in a set of points unchanged under these involutions, the equations of C_1 are

$$x:y:z:w = A\,(t^4 - 1):B\,(t^3 + t):C\,(t^3 - t):D\,(t^4 + 6\mu t^2 + 1), \quad 9\cdot22$$

where A, B, C, D are determined by the choice of unit point; we use $A = D = 1$, $B = C = 2$. The fundamental equation is

$$t^4 + 6mt^2 + 1 = 0,$$

where $3m\mu + 1 = 0$. The equation of the unique quadric (h) containing C_1 is

$$(h): \quad x^2 + \left(\frac{1 + 3\mu}{2}\right)^2 y^2 - \left(\frac{1 - 3\mu}{2}\right)^2 z^2 = w^2. \qquad 9\cdot23$$

The equations of C_1 given above are obtained from [4] by projecting C from the point $(1, 0, m, 0, 1)$ upon any prime $\Sigma\xi_i x_i = 0$, the transformation from the osculating system of reference being

$$x = A\,(x_4 - x_0), \quad y = -B\,(x_3 + x_1), \quad z = -C\,(x_3 - x_1),$$
$$w = D\,(x_4 + x_0 + 6\mu x_2), \quad u = \Sigma\xi_i x_i = 0.$$

9·2,1. The principal tetrahedron and the tetrahedron whose vertices are given by $t^4 + 6pt^2 + 1 = 0$ on C_1 are in four-fold perspective, the vertices of perspection being a similar group of four points on the curve which is symmetrical with C_1 in regard to its centre. This curve has the same principal tetrahedron and lies on the same quadric as C_1; it meets C_1 on the principal axes and on the principal plane.

2. A principal axis of C_1 is an axis of its developable, being the intersection of the osculating planes at the points where it meets C_1. The principal axes are the only "chord-axes" of C_1.

---

* This naturally does not apply to a curve or surface projectively related to C_1 *and* some other elements which are not themselves symmetrical, e.g. the surface generated by the chords of an involution on C_1 if this involution is not the principal involution.

3. Through a point on a principal directrix there is one chord of the corresponding principal involution; through a point on the principal axis there are two.

4. The tangents at the ends of a principal chord meet the corresponding directrix at points harmonic to the points where the directrix meets the other two principal axes.

5. The centre of C_1 is the trace of the g-line through the vertex of projection of C, and the principal axes are the traces of the director planes through it. The principal directrices are the projections of the triangle of director lines lying in its polar plane.

6. The principal plane meets the curve in a set of the involution given by $t^4 + 6pt^2 + 1 = 0$; this set has the principal directrices as diagonal triangle, and is the only coplanar set of this involution.

7. The rational quartic curve in the plane is the projection of C from a line. In this case there is a fundamental quartic involution corresponding to the points on the line of projection. There are two linearly independent conditions that four points should be collinear. Each of the three nodes is given by a quadratic apolar to some member of the involution.

The flexes

9·3. When $t_1 = t_2 = t_3 = t_4 \,(=t)$ in $(ft_1)(ft_2)(ft_3)(ft_4) = 0$, this condition for coplanar points becomes $(ft)^4 = 0$: hence at each of the four points on C_1 given by the f-equation, the osculating plane is stationary, meeting C_1 at four coincident points. Such a point will be called a *flex* of the curve; it is the analogue of a point on a plane curve at which there is a stationary tangent.* The stationary osculating planes (or *flex planes*) are the traces of the osculating primes of C which pass through the vertex of projection.

The four flexes do not lie in one plane unless $(ff')^4 = 0$, i.e. the invariant i of the f-quartic vanishes (2·43).

The trisecants and hessian points

9·4. If the condition $(ft_1)(ft_2)(ft_3)(fT) = 0$ holds for all values of T, the points t_1, t_2, t_3 are in line: they are a triad of points on a trisecant of C_1. The cubic involution (g_3^1) cut out by the trisecants is therefore given by the pencil of cubics apolar to the f-quartic.

---

* In general there is no stationary tangent to C_1.

The g_3^1 has four double points, at each of which a trisecant touches the curve, these points t being given by $(ft_1)(ft)^2(fT) = 0$, or, on eliminating t_1, by $(ff')^2(ft)^2(f't)^2 = 0$. Hence the points of contact of the tangent trisecants are given by the hessian of the f-quartic; they are called the *hessian points* of C_1. (Cf. 3·11.)

The flexes and hessian points each form tetrads symmetrical in the axes: for the canonical form 9·22 they are given by

$$f \equiv t^4 + 6mt^2 + 1 = 0, \quad h \equiv mt^4 + (1 - 3m^2)t + m = 0.$$

Every symmetrical tetrad of points on C_1 is a set of the involution, $f + \lambda h = 0$, determined by these two. The jacobian of the involution, which is the sextic covariant $(gt)^6$ of the f-quartic, is the product of three mutually harmonic quadratics determining the ends of the principal chords. (Cf. 6·2 (ii).)

9·4, 1. The points where the tangent trisecants meet C_1 again are given by
$$4m^3t^4 + (1 - 6m^2 - 3m^4)t^2 + 4m^3 = 0,$$
that is, by $jf - ih = 0$; this is known as the *steinerian* of f.

The hessian points are coplanar only if $j = 0$; the steinerian points are coplanar if either $i = 0$ or $j = 0$.

2. If $(ft)^3(fT) = 0$, two roots of the f-quartic are equal and C_1 has a stationary tangent. If $(ft_1)(ft_2)(fT)^2 = 0$, the invariant j of the f-quartic vanishes: f has a quadratic apolar to it and C_1 has a node. If $(ft)^2(fT)^2 = 0$, the invariants i and j both vanish and C_1 has a cusp.

Cubic involutions on C_1: cubic osculants

9·5. With $t_1 = t_2 = t_3 = t$ the condition for coplanar points is $(ft_0)(ft)^3 = 0$: thus through any point t_0 of C_1 pass three planes osculating at a triad of points t given by the first polar of t_0 with respect to the f-quartic. Further, the plane of the triad passes through t_0, since $(ft_0)(f't_0)(ff')^3 \equiv 0$. With the equations 9·22 this plane is found to be
$$4t_0 x + (3m + 1)(1 - t_0^2)y + (3m - 1)(1 + t_0^2)z = 0,$$
and when t_0 varies, the envelope of the plane is the quadric cone

$$(q): \quad 4x^2 + (3m + 1)^2 y^2 - (3m - 1)^2 z^2 = 0.$$

Thus the first polars of the f-quartic describe a g_3^1 of sets of points whose planes envelope the quadric cone (q). This cone has the principal chords for axes and touches C_1 at each of the flexes, where t_0 coincides with t. At a double point of the involution two of the osculating planes from t_0 coincide and $\overline{t_0 t}$ is a trisecant touching at t. Thus the double points of the g_3^1 of first polars of f are the same as those of the g_3^1 of trisecants, namely the hessian points of C_1. We shew this to be a particular case of a more general proposition concerning cubic involutions on C_1.

In [4], any cubic involution on C, being determined by two of its members, is cut out by the trisecant planes through a point (x) and these planes form one system of generating planes of a quadric cone. The double points are given by the roots of the hessian of the quartic represented by (x); hence when the double points are given there are two possible involutions and thus two points (x) and (y), these being conjugate on a g-line. The one involution may be written either as

$$(xt_1)(xt_2)(xt_3)(xT) = 0, \quad \text{or} \quad (yt_0)(yt)^3 = 0,$$

while the other is given by interchanging x and y. The two are connected by the relations

$$(xy)^3 (xT_1)(yT_2) \equiv 0.$$

On projection it appears that in general *the planes of the triads of a g_3^1 on C_1 envelope a quadric cone*, the exceptional case being when (x) is the vertex of projection and we have the trisecants of C_1 lying on the quadric (h). *To any g_3^1 on C_1 there is a conjugate g_3^1*; every triad of the one is apolar to every triad of the other, and the two involutions have the same double points. These lead to a pair of complementary quadric cones. In particular the involutions of trisecants and of first polars of points of C_1 with respect to the f-quartic are conjugate. Here is another case, arising when (x) and (y) are collinear with the vertex of projection. A quadric cone which touches C_1 at a symmetrical set of points has its vertex at the centre of C_1 and is itself symmetrical in regard to the principal axes of C_1. If the points of contact of one such cone are $f + \lambda_1 h = 0$, then the first polars of $f + \lambda_1 h$ determine planes enveloping another such cone touching C_1 at $f + \lambda_2 h = 0$, where $f + \lambda_1 h$ and $f + \lambda_2 h$ are apolar. The common tangent planes of the two cones touch C_1 at the points given by the hessian of $f + \lambda_1 h$ or $f + \lambda_2 h$. The planes enveloping the cone which touches at the tetrad on the principal plane pass through the trisecants, this cone being the enveloping cone of the quadric containing C_1.

The interpretation of first polars in regard to the f-quartic can be manipulated so as to extend to first polars of any quartic $(\alpha t)^4$. The first polar of t_0 on C in [4] defines the points of contact of osculating planes of C which meet the line joining the points (f) and t_0. Now take a general point (x) instead of (f), and project from t_0 instead of (f). The point representing $(xt)^4 = 0$ is projected into a point representing $(xt)^3 (xt_0) = 0$ with reference to the normal cubic curve into which C is projected. To link this result with C_1 by projection from (f) we consider the dual: the osculating prime of C at t_0 meets the prime representing $(\xi t)^4 = 0$ in a plane representing $(\xi t)^3 (\xi t_0) = 0$ with reference to the osculant of C at t_0. Moreover, the *osculant of C_1 at a point t_0* of C_1 is defined as the projection of the osculant at t_0 of C: it is a twisted cubic (unless t_0 is a flex) which meets all the tangents of C_1 and has at t_0 the same tangent and osculating plane as C_1. Our interpretation of first polars of $(\alpha t)^4$ may therefore be expressed as follows: *any plane meeting C_1 in $(\alpha t)^4 = 0$ meets the osculant at t_0 in $(\alpha t)^3 (\alpha t_0) = 0$.* The equations of the osculant of C_1 (9·11) are thus $x_i = (\alpha_i t)^3 (\alpha_i t_0)$; and as the point t on the osculant is on the tangent to C_1 at t, the same equations are also the equations of the tangent at t when t_0 is variable, as may readily be seen directly. When t and t_0 are both variable the equations give the points of the tangent surface of C_1.

9·5, 1. Every osculant of C_1 has the flex planes as osculating planes, and the cross-ratio of these planes on each osculant is the same as the cross-ratio of the four flexes on C_1.

2. Any axis of any osculant meets the flex planes in the same cross-ratio.

3. The fundamental linear complexes of the osculants belong to a linear ∞^2 system, the common lines forming a regulus containing the tangents to C_1 at its flexes.

4. A plane osculating an osculant meets C_1 in four harmonic points.

Osculating planes and the e-polarity

10·1. The planes which meet C_1 in equianharmonic tetrads, i.e. tetrads for which $i = 0$, envelope a quadric (e) whose class equation referred to the principal tetrahedron is

$$(e): \qquad \xi^2 + \eta^2 - \zeta^2 = 3\,(1 + 3\mu^2)\,\omega^2;$$

while those planes which meet C_1 in harmonic tetrads $(j = 0)$ envelope a class cubic surface (k) given by

$$(k): \qquad \mu\xi^2 + \tfrac{1}{2}\,(1-\mu)\,\eta^2 + \tfrac{1}{2}\,(1+\mu)\,\zeta^2 - \mu\,(1-\mu^2)\,\omega^2 = \xi\eta\zeta/\omega.$$

Any osculating plane of C_1 meets the curve in a tetrad for which both $i = 0$ and $j = 0$. Thus the osculating developable of C_1 is the sextic developable common to the quadric and cubic envelopes (e) and (k). It is the projection of the tangent surface of C (2·6).

In [4] the primes meeting C in equianharmonic and harmonic sets are those which touch the primal I and the surface K respectively. On projecting, the quadric (e) and the class cubic surface (k) appear as the section of the enveloping cone to I and the projection of K respectively. The curve C_1 thus lies upon (k), the osculating planes touching (k) at points of C_1; in other words, C_1 *is asymptotic on* (k).

From the fundamental polarity in [4] we are now led to consider the polarity in regard to the quadric (e). It makes a slight simplification of language if we suppose projection to be made on the polar prime Φ of the vertex of projection F. Then the quadric (e) is the section of I, and the e-polar of a point is the section of its I-polar prime. The section of the elements of the complete figure of C is the e-reciprocal of the projection of the dual elements.* Thus the osculating developable of C_1 is the e-reciprocal of the section of the tangent surface of C—a sextic curve, E^6, which is the curve along which the osculating planes (and hence tangents) of C_1 touch (e). In view of the e-polarity, this curve has four cusps at the flexes of C_1, ∞^1 triaxes on a quadric corresponding to the trisecants of C_1 on (h), four of these triaxes being tangent to the

---

* With the phraseology of 2·6 the complete figure in [4] has characteristic numbers (4, 6, 6, 4), so that those of projection and section of its elements are (4, 6, 6) and (6, 6, 4) respectively.

curve; and so on. It is the intersection of (e) with the cubic surface (J, Φ) which is the e-reciprocal of (k); hence every generator of (e) is a trisecant of E^6 and thus a triaxis of C_1.

Also, as a result of our former representation in [4], any quartic equation $(at)^4 = 0$ with $(af)^4 = 0$ is represented by a point or a plane in [3], the point and plane being pole and polar with respect to (e). The plane $(\alpha t)^4 = 0$ passes through the point $(at)^4 = 0$ if $(a\alpha)^4 = 0$; consequently, if t_i are the points of C_1 (9·11) on the plane, the coordinates of the pole of this plane with respect to (e) are

$$x_i = (\alpha_i t_1)(\alpha_i t_2)(\alpha_i t_3)(\alpha_i t_4).$$

10·1,1. Each flex plane touches (e) at the flex.

2. The point of contact of an osculating plane with (e) lies on the cubic osculant of the point in which the plane meets C_1 again.

3. The g-lines in Φ are the trisecants of the curve $(K, \Phi) \equiv C^\star$, and form a regulus on a quadric (g). The g-planes meet Φ in the lines of a regulus on a quadric (f). The planes through any line of this regulus cut C_1 in sets of a syzygetic pencil; in particular three bitangent planes of C_1 pass through the line. The regulus contains the tangents at the four flexes.

The tangents

11·1. Six tangents of C_1 meet an arbitrary line l, their contacts being given by the jacobian of the g_4^1 cut out by planes through l.

The tangents at a symmetrical tetrad of points on C_1 are themselves a symmetrical set and therefore their line coordinates are linearly connected; hence they belong to a regulus on a symmetrical quadric. They are the projection of a tetrad of tangents of C in [4] for which the fifth associated line is the g-line through the vertex of projection. Two of these quadrics are the quadric (h) of trisecants of C_1 and the quadric (f) of triaxes of the bitangent developable of C_1 (10·1, 3).

The tangents of C_1 belong to a tetrahedral complex, every tangent meeting the four flex planes in the cross-ratio of the four flexes on C_1, as may be verified by referring the curve to its flex planes. It is seen more directly from a general theorem concerning the normal curve C. The cross-ratio of the primes joining any trisecant plane of C to four fixed points on C is constant, and dually, any tangent to C meets four fixed osculating primes in a constant cross-ratio; the theorem in [3] follows by projecting from the point common to the four primes. The complex is in fact that of the axes of the osculants $(9·5, 2)$.

If the tangents are represented in the usual way by points of a quadric Ω in [5], the points lie on a rational curve of order six, C^6. This curve, being the projection of the normal sextic from a point, has a fundamental sextic equation, and as any collineation $C_1 \rightarrow \bar{C}_1$ induces a collineation $C^6 \rightarrow \bar{C}^6$, this sextic equation must be $(gt)^6 = 0$, where $(gt)^6$ is the sextic covariant of f. The quartics apolar to $(gt)^6$, which are $f + \lambda h$, determine the quadrisecant planes of C^6; these planes meet Ω in conics representing the reguli to which the symmetrical tetrads of tangents on C_1 belong. *The geometry of the tangents of C_1 is the geometry of the points of this C^6 in* [5]. Six tangents of C_1 belong to an arbitrary linear complex; the contacts of these tangents are given by a sextic apolar to $(gt)^6$. Similarly, any one of the ∞^3 quintics apolar to $(gt)^6$ gives a set of five tangents having two lines meeting them.

The flexes of C^6 in [5] correspond to the ends of the principal chords on C_1, and the four points on C^6 corresponding to the flexes on C_1 are the points where the osculating planes are generating planes of Ω. The tetrahedral complex containing the tangents of C_1 is represented by the intersection of Ω with another quadric through C^6 having these same generating planes (cf. projection of C^6 on G from a point of G; $6·5, 3-5$).

Quadratic involutions on C_1

12·1. Chords of C_1 which meet a given chord p meet the curve in pairs of points of an involution (g_2^1). If the ends of the chord p are given by $(pt)^2 = 0$, then $(fp)^2 (ft_1) (ft_2) = 0$ is the involution and $(fp)^2 (ft)^2 = 0$ its double points. The double points are at the ends of p only when p is a principal chord and the involution is a principal involution $(9·2, 2)$.

Any g_2^1 on C_1 can be so constructed from a chord p, for if two members of the g_2^1 are taken there is one chord meeting the two chords corresponding to these members. In particular there is one chord through the intersection of two chords which meet, and hence there are three chords of C_1 through an arbitrary point.

Consider the g_2^1 on C_1 arising from the chords meeting the chord p. Through any point of p there are two chords of the involution each determining a plane through p. These pairs of planes are themselves in an involution of which the double planes $X = 0$, $Y = 0$ meet C_1 in pairs $(qt)^2 = 0$ and $(rt)^2 = 0$. This involution of planes is then $X^2 = \lambda Y^2$, meeting C_1 in

$$[(qt)^2]^2 = \lambda [(rt)^2]^2,$$

which is the pencil of quartics cut out by a pencil of planes $Z = \lambda W$ through a line l; two of the planes through l, $Z = 0$ and $W = 0$, are bitangent to C_1 at $(qt)^2 = 0$, $(rt)^2 = 0$ respectively. *The chords of the involution are thus generators of the cubic scroll* $X^2 W = Y^2 Z$, on which the lines $X = Z = 0$ and $Y = W = 0$ are torsal lines or double generators, while the chord p, $X = Y = 0$, is the nodal line and l, $Z = W = 0$, is the directrix.

The formation of this scroll is more readily seen on projecting the normal cubic scroll which we know to be the locus of chords of a g_2^1 on C in [4] (4·11). There the chords meet a director line $(x_0 = x_2 = x_4 = 0)$ and a director plane $(x_1 = x_3 = 0)$. The director line projects into the directrix of the scroll in [3], and if the director plane passes through the vertex of projection it gives a principal chord of C_1, which is in this case the nodal line, the g_2^1 being one of the three associated principal involutions. If on the other hand the director plane does not pass through the vertex, the projected involution is not induced by a collineation in [3] and the nodal line does not join the double points. In [4] a prime through a generator $\dfrac{x_0}{x_2} = \dfrac{x_1}{x_3} = \dfrac{x_2}{x_4} = \lambda$ meets the cubic scroll in a conic whose plane has equations of the form

$$\lambda_0 x_0 + \lambda_1 x_1 + \lambda_2 x_2 = 0,$$
$$\lambda_0 x_2 + \lambda_1 x_3 + \lambda_2 x_4 = 0,$$

and the primes through any one of these planes meet the scroll again in the generators. These planes are called the conic-planes for the scroll. There is just one conic-plane through an arbitrary point, and thus arises the nodal line in [3], which is double in that each point P of it stands for two points of the conic on the conic-plane; also the pairs of points corresponding to the points P form an involution which is projective with the range of the points P, the double elements of this involution corresponding to the torsal elements.

12·1,1. If the double points of the involution are $t=0$ and $t=\infty$, the contacts of the torsal elements are given by $f_0 t^4 - 2f_2 t^2 + f_4 = 0$ and the ends of the nodal chord by

$$(f_0 f_3 - f_1 f_2)\, t^2 - (f_0 f_4 - f_2^2)\, t + f_1 f_4 - f_2 f_3 = 0.$$

2. The directrix of the cubic scroll of involution chords cannot be a chord of C_1 except in the case when it coincides with the nodal line; this occurs when the tangents at $(pt)^2 = 0$ intersect, and the cubic scroll is then of the Cayley form.

3. If a rational quartic curve C_1 lie on a cubic scroll so that the generators of the scroll each meet the curve twice, then they are the chords of an involution on the curve. If the curve lie on the scroll so that the generators meet it only once, then the directrix is a chord and the nodal line is a trisecant of the curve.

4. A rational quartic curve C_1 may be regarded as the intersection of a quadric and a cubic scroll having a trisecant as nodal line.

5. If the tangent planes of a quadric cone correspond projectively to the generators of a regulus on a quadric, then the locus of the intersection of corresponding elements is in general a rational quartic curve.*

6. If the planes of a pencil correspond projectively to pairs of generators in an involution on a quadric regulus, then the locus of the intersection of corresponding elements is a rational quartic curve.*

7. A variable line of a regulus on a quadric (h), meets three fixed planes at P_1, P_2, P_3. If on the line a fourth point P is taken so that the cross-ratio (P_1, P_2, P_3, P) remains constant, then the locus of P is a rational quartic curve.

Conversely, the generators of the quadric (h) which each meet C_1 in a single point P meet the planes formed by any three fixed concurrent chords at P_1, P_2, P_3, so that (P_1, P_2, P_3, P) is constant.†

8. A prime section of the surface K is a quartic curve C^*. An involution on C^* is given by conics of K passing through a fixed point O. The directrix is given by the tangent plane to K at O and the nodal line by the conic-plane of K having O for isolated point.

---

* Deaux, *Atti del Congresso Internazionale*, 4 (1928) 185.
† Vietoris, *Wien. Ber.* 125 (1916) 259.

Bitangent planes: the nodal curve and g-polarity

13·1. With $t_1 = t_3$ and $t_2 = t_4$, the coplanarity condition becomes $(ft_1)^2 (ft_2)^2 = 0$, expressing that there is a plane bitangent to C_1 at t_1, t_2 or that the tangents at t_1, t_2 intersect. The bitangent planes form a developable (the *bitangent developable* of C_1) and the point of intersection of the tangents traces out a curve which is the nodal curve of the osculating developable of C_1 (the *nodal curve of C_1*). The e-reciprocal of the bitangent developable of C_1 is the nodal curve of E^6 (10·1): this nodal curve is immediately seen to be the Φ section of the surface K which is the locus of the intersection of osculating planes at t_1, t_2 on C. It is therefore a rational quartic curve, $C^\star$, whose trisecants are the g-lines in Φ. On reciprocation it follows that *there are four bitangent planes of C_1 through an arbitrary point, the complete figure being* (6, 6, 4); and that all bitangent planes of C_1 touch a quadric (f); they pass by threes through the lines of one regulus of (f) and singly through the lines of the other.

The two-two correspondence given by $(ft_1)^2 (ft_2)^2 = 0$ indicates that there are two tangents of C_1 meeting any given tangent; these are coincident when $(ff')^2 (ft_1)^2 (f't_1)^2 = 0$ and then t_1 is one of the hessian points of C_1. The two tangents coincide with the given one when $(ft_1)^4 = 0$ and then t_1 is one of the flexes. Hence *the nodal curve of C_1 passes through the flexes and touches the tangent trisecants at the steinerian points of C_1* (9·4, 1).

The question arises as to whether some three tangents of C_1 meet at a point. The conditions for this to occur are

$$(ft_1)^2 (ft_2)^2 = (ft_2)^2 (ft_3)^2 = (ft_3)^2 (ft_1)^2 = 0,$$

in which case t_1, t_2 are the roots of the equation $(ft_3)^2 (ft)^2 = 0$, and

$$t_1 t_2 : t_1 + t_2 : 1 = (ft_3)^2 f^2 : 2 (ft_3)^2 f : (ft_3)^2.$$

Substituting in $(ft_1)^2 (ft_2)^2 = 0$, we have $(ft_3)^4 (ff')^4 = 0$. So, neglecting the flexes for which the three tangents are coincident, we find that *three concurrent tangents can only occur when C_1 is equianharmonic*; and then there are ∞^1 triads of concurrent tangents. The two-two correspondence is here formed by the g_3^1 of the contacts of the triads of tangents as in the case of Poncelet's porism for triangles inscribed in one conic and circumscribed to another in the plane (cf. 5·1,3).

Similarly, if there is a skew quadrilateral of tangents, i.e. a chain of four tangents each meeting the next so that the conditions are

$$(ft_1)^2 (ft_2)^2 = (ft_2)^2 (ft_3)^2 = (ft_3)^2 (ft_4)^2 = (ft_4)^2 (ft_1)^2 = 0,$$

then $(ft_1)^2 (ft)^2 = 0$ and $(ft_3)^2 (ft)^2 = 0$ are the same equation; and likewise $(ft_2)^2 (ft)^2 = 0$ and $(ft_4)^2 (ft)^2 = 0$. Hence there is a quadratic $(pt)^2$ harmonic to each of these quadratics, and $(fp)^2 (fT)^2 = 0$ for all values of T. The f-quartic in this case must be harmonic and C_1 has ∞^1 skew quadrilaterals of tangents. The f-equation may be written $t^4 - 1 = 0$ and the two-two correspondence is $t_1^2 t_2^2 = 1$, breaking up into two quadratic involutions $t_1 t_2 = \pm 1$; the points of contact of the four tangents form a set of a syzygetic involution on the curve.

When $(ft_1)^2 (ft_2)^2 = 0$ the osculating planes at t_1, t_2 on C meet in a point P of $C^\star$ and the projected osculating planes meet in an axis l which passes through P and the point of intersection of the tangents. By taking a consecutive position it is seen that the chord $t_1 t_2$ is a line of the bitangent developable (the e-reciprocal of the tangent to $C^\star$ at P) and that the axis l is the tangent to the nodal curve. We prove that l touches the unique quadric (g) upon which $C^\star$ lies (l being the g-reciprocal of the tangent to $C^\star$ at P), from which it follows that *the nodal curve is* the g-reciprocal of the osculating developable of $C^\star$ and that it is therefore *a rational sextic curve of the same type* (6, 6, 4) *as, and indeed projectively equivalent to,* E^6 *which is the e-reciprocal of C_1.* The result required is simply verified by taking the osculating planes to C as $x_0 = x_1 = 0$, $x_3 = x_4 = 0$, so that for Φ to pass through P $(0, 0, 1, 0, 0)$, the vertex of projection F must lie in $x_2 = 0$ and have coordinates $(f_0, f_1, 0, f_3, f_4)$. Then any point Q of the axis l in which the projected osculating planes meet is $(f_0, f_1, \lambda, -f_3, -f_4)$. When Q lies on (g) the hessian point of Q lies on the g-line in Φ through it. But the hessian point of Q only lies in Φ when λ is infinite and then Q is at P. Hence the two g-lines in Φ which are met by l coincide, and l touches the quadric (g) at P.*

13·1,1. The g-line meeting the axis l is $(f_0, 0, \mu, 0, f_4)$ and the tangent of $C^\star$ is $(0, f_1, \nu, -f_3, 0)$; the axis and tangent are harmonic with respect to the generators through P of the quadric (g).

2. The J-polar quadric of the vertex of projection F meets the surface K in $C^\star$ and C, and thus meets Φ in $C^\star$. The chords of C which lie in the quadric project into the lines of the bitangent developable of C_1.

3. If Q, $(qt)^4 = 0$, is a point of an axis l, and P, $(pt)^4 = 0$, is the point of $C^\star$ on this axis, then
$$(pq)^2 (pf)^2 (qf)^2 = 0.$$

4. The four cusps of the nodal curve are at the steinerian points of C_1, and the hessian points of the cusps at the flexes of C_1. The osculating planes at the cusps are the osculating planes at the hessian points of C_1.

5. By means of the tangents of C_1, a two-two correspondence is set up between the points of C_1 and the points of its nodal curve; the cross-ratio of the four cusps on the nodal curve is the cross-ratio of the four flexes on C_1.

6. The tangents to C_1 each meet the nodal curve at a pair of points t_1, t_2 of a two-two correspondence given by $(ft_1)^2 (ft_2)^2 = 0$.

7. The nodal curve of the bitangent developable of C_1 is C_1; the bitangent developable of the nodal curve of C_1 is the osculating developable of C_1.

8. Through a point of the nodal curve of C_1 there is one chord of C_1, and two further osculating planes neither passing through l but meeting in an axis m; the chord and the two axes l and m lie in one plane.

The Steiner surface (k)

14·1. We now extend the interpretation of second polars as we have that of first polars, and examine a further osculant. The second or *conic osculant* is the osculant of the first or cubic osculant. Thus in [4] the second osculant of C at t_0 is the trace of the osculating planes of C on the osculating plane at t_0 and is the conic of K on this plane. The conic osculant of C_1 is the projection of this conic and we may regard the surface (k) which is the projection of K as the locus of such osculants.

---

* For a more comprehensive account v. Telling, *Proc. Camb. Phil. Soc.* 29 (1933) 195.

A plane meeting C_1 where $(\alpha t)^4 = 0$ meets the osculating plane t_0 in a line represented by $(\alpha t)^2 (\alpha t_0)^2 = 0$ with reference to the osculant. In particular each flex plane touches the osculant. Thus the equations of the second osculant of 9·11 at t_0 are $x_i = (\alpha_i t)^2 (\alpha_i t_0)^2$; and these equations are also those of the second osculant at t when t_0 is variable. When t and t_0 are both variable the equations are those of the surface (k), each point of (k) being given by the parameters of the two osculants through it. The corresponding point of K was mapped (6·1) on a plane π so that t, t_0 were the parameters of the tangents to a reference conic C_0 on π; and the projected surface (k) can accordingly be mapped on π as follows.

A plane meets the surface (k) in a rational quartic curve whose equation is $(\alpha t)^2 (\alpha t_0)^2 = 0$; this is represented on the plane π by $(\alpha x)^2 (\alpha x')^2 = 0$, a conic outpolar to C_0 and meeting it at $(\alpha t)^4 = 0$. Since $(\alpha f)^4 = 0$, this conic is also outpolar to $(f\xi)^2 (f\xi')^2 = 0$. Thus plane sections of (k) are represented on π by conics outpolar to two and therefore to a range of conics which may be called the *basal range*. If $(\alpha t)^4$ is harmonic, the correspondence $(\alpha t)^2 (\alpha t_0)^2 = 0$ breaks up (13·1) into two involutions, $(At)(At_0) = 0$ and $(Bt)(Bt_0) = 0$ with $(AB)^2 = 0$; thus any plane meeting C_1 in harmonic points meets (k) in two conics; and as the two involutions have one common pair, the plane touches (k) at that one intersection of the two conics which is given by the quadratic apolar to $(\alpha t)^4$. The two conics are represented on π by two lines conjugate to the basal range and this one intersection of the two conics by the intersection of the two lines. The other three intersections of the two conics, as also the three nodes of any plane section, are due to three nodal lines of (k) which, being the projection of the conics of K in the three director planes through the vertex, are the principal axes of C_1. A point on a nodal line is the projection of two points upon the conic of K and is thus mapped on π by two distinct points of one side of the common self-conjugate triangle of the basal range and conjugate in regard to the range; the involution of such pairs of points will be referred to as a *basal involution*. The centre of C_1 is a triple point of (k) and is represented on π by the three vertices of the triangle. Save

for these exceptional points the mapping is a unique correspondence between the points of (k) and the points of π.

The surface of Veronese $K^\star$

14·2. The surface (k) is the general surface in [3] which can
be mapped on a plane so that its plane sections correspond to
conics on the plane. It is known as the Quartic Surface of Steiner.
The surface which is mapped point for point on a plane π so that
its prime sections correspond to the conics of π is a surface $K^\star$ in
[5] known as the surface of Veronese. As two conics on π meet in
four points it is a surface of order four; its equations may be taken
to be

$$x_0 : x_1 : x_2 : x_3 : x_4 : x_5 = u_1^2 : u_2^2 : u_3^2 : 2u_2 u_3 : 2u_3 u_1 : 2u_1 u_2, \quad 14\cdot21$$

so that the coordinates of any prime in [5] are the coefficients of
the conic of π corresponding to the prime section. The primes
which pass through a point P of [5] correspond to conics whose
coefficients satisfy a linear relation; hence the projection of $K^\star$
from P is a surface in [4] whose prime sections are mapped into
conics of π which are outpolar to a fixed conic. This is the surface
K. The unique quartic curve C to which K belongs is the projection
of the curve on $K^\star$ whose points correspond to the points of the
fixed conic; since these lie on a prime not passing through P we
may take this as the prime ϖ upon which we project from P and
thus say that K and C are the projection and section of $K^\star$ on ϖ.

The primes which pass through a line l of [5] correspond to
conics on π which are outpolar to a fixed range of conics. Thus the
projection of $K^\star$ from a line is a surface (k), projection from l
being equivalent to successive projection from two points of l.
From P on the line l we project onto the prime ϖ and then from
Q, the point in which l meets ϖ, onto a solid: the normal curve C
which is the section of $K^\star$ by ϖ is projected from Q into C_1 on (k).
The points P and Q are effectively equivalent to two conic
envelopes of the basal range on π.

Lines on π represent conics on the surface of Veronese and its
projections; the osculant conics are the projections of those which
touch the section ϖ. Thus on (k) there are ∞^1 series of conic

osculants, each series being osculant to a particular quartic curve $C_{(\mu)}$ whose points are represented by the points of a conic of the basal range on π. This curve $C_{(\mu)}$ is asymptotic on (k): for, the tangent to $C_{(\mu)}$ touches two consecutive osculants of the series and thus meets (k) at three consecutive points. Hence the conics of the range on π correspond to the asymptotic curves on (k), two curves passing through each point. There are four osculants which belong to every series: on π they are represented by the four lines touching the conics of the basal range; and since these lines are self-conjugate for every conic of the range, the plane of each of these osculants touches (k) at every point of the osculant. These are the osculants at the flexes of all the curves $C_{(\mu)}$; any conic of (k) touches the flex planes at points of these *singular* conics. The flex planes are the *tropes* of (k).

14·2,1. In general each conic of (k) is osculant to one asymptotic curve.

2. The quartic curve corresponding to $(f\xi)^2 (f\xi')^2 = 0$ has for conic osculants those conics of (k) which meet C_1 in the points of contact of bitangent planes; it passes through the hessian points of C_1.

3. The quartic curve corresponding to $(fx)^2 (fx')^2 = 0$ is $C^\star$ whose points are the reciprocal with respect to (e) of the bitangent planes of C_1. The hessian points of $C^\star$ are the flexes of C_1.

4. The osculants of C_1 which touch $C^\star$ do so at the flexes of $C^\star$ and the cross-ratio on $C^\star$ of its flexes is equal to the cross-ratio on C_1 of its four flexes; C_1 and $C^\star$ having the same invariant are projectively equivalent, and so also are the bitangent and osculating developables of C_1.

5. Given any twisted quartic curve on (k), there is one asymptotic quartic $C_{(\mu)}$ which passes through its four hessian points. The osculants of $C_{(\mu)}$ which touch the given quartic do so at the flexes.

6. The points of a quartic curve C_1 may be mapped on the points of a conic (ϖ), and their geometry discussed by means of an auxiliary conic envelope (Q) which is inpolar to (ϖ). The following elements of C_1 and (ϖ) correspond: four coplanar points—a polar tetrad with respect to (Q); three points of a trisecant—a polar triad; the flexes —the points on (ϖ) of contact of common tangents of (ϖ) and (Q); the hessian points—the points of (Q) on (ϖ); the principal chords— the sides of the common self-polar triangle of (Q) and (ϖ); etc. The

conics (ϖ) and (Q) may be taken as $(xx')^2 = 0$, and $(f\xi)^2 (f\xi')^2 = 0$, where $(ft)^4$ is the f-quartic of C_1.*

7. The projection of the surface of Veronese given by 14·21, from the line $(p, q, r, 0, 0, 0)$, where $p + q + r = 0$, on to the solid $x_1 = x_2 = 0$ is

$$u_1^2 + u_2^2 + u_3^2 : 2u_2 u_3 : 2u_3 u_1 : 2u_2 u_3.$$

Equations of (k) and of its asymptotic curves

14·3. Referred to the principal tetrahedron, the equations of $C_{(\mu)}$ are

$$x = A\,(t^4 - 1), \quad y = 2B\,(t^3 + t), \quad z = 2C\,(t^3 - t), \quad w = t^4 + 6\mu t^2 + 1.$$

Its conic osculants are

$$x = 2Av_2 v_3, \quad y = 2Bv_3 v_1, \quad z = 2Cv_1 v_2, \quad w = a^2 v_1^2 + b^2 v_2^2 + c^2 v_3^2,$$

where

$$v_1 = t + t_0, \quad v_2 = tt_0 + 1, \quad v_3 = tt_0 - 1$$

and

$$a^2 = 2\mu, \quad b^2 = 1 + \mu, \quad c^2 = 1 - \mu.$$

The unit plane of the reference system may be taken to be one of the flex planes, so that $x + y + z + w = 0$ meets (k) in a repeated osculant. Then $A^2 = b^2 c^2$, $B^2 = c^2 a^2$, $C^2 = a^2 b^2$, and the possible variations of sign being covered by the symmetry, the parametric equations of (k) are

$$x = 2u_2 u_3, \quad y = 2u_3 u_1, \quad z = 2u_1 u_2, \quad w = u_1^2 + u_2^2 + u_3^2, \quad 14\cdot31$$

where $u_1 = av_1$, $u_2 = bv_2$, $u_3 = cv_3$ give the series of osculants depending on the one parameter μ, while the asymptotic curves are obtained on putting $t_0 = t$. If u_1, u_2, u_3 are point coordinates in the plane π, the equations 14·31 express the mapping of (k) upon π, the basal range being $pv_1^2 + qv_2^2 + rv_3^2 = 0$ with $p + q + r = 0$.

14·3,1. The locus and envelope equations of (k) are respectively

$$y^2 z^2 + z^2 x^2 + x^2 y^2 = 2xyzw \quad \text{and} \quad (\xi^2 + \eta^2 + \zeta^2 - \omega^2)\,\omega = 2\xi\eta\zeta.$$

The latter may be put in the form $\sigma\omega = 2\xi_1 \eta_1 \zeta_1$, where $\sigma = 0$ is one of the singular conics and $\xi = \xi_1 + \omega$, $\eta = \eta_1 + \omega$, $\zeta = \zeta_1 + \omega$.

2. The four singular conics of (k) lie on the quadric $x^2 + y^2 + z^2 = w^2$, and touch in pairs on the nodal lines.

3. Referred to the tetrahedron of flex planes with the centre as unit point, the equations of the osculants of C_1 are

$$x_i = \frac{(\alpha_i t)^2 (\alpha_i t_0)^2}{(\alpha_i \alpha_l)^2 (\alpha_i \alpha_m)^2 (\alpha_i \alpha_n)^2}, \quad \text{where } l, m, n, i \text{ are distinct, and } \alpha_i \text{ are}$$

the parameters of the flexes.

* Weyr, *Wien. Ber.* 72 (1875) 686.

The osculant at the flex α_0 is

$$x_0 = 0, \quad x_1 = (\alpha_1 t)^2 (\alpha_2 \alpha_3)^2, \quad x_2 = (\alpha_2 t)^2 (\alpha_3 \alpha_1)^2, \quad x_3 = (\alpha_3 t)^2 (\alpha_1 \alpha_2)^2.$$

The point equation of (k) is

$$\sqrt{x_0} + \sqrt{x_1} + \sqrt{x_2} + \sqrt{x_3} = 0,$$

and its envelope equation is

$$\xi_0^{-1} + \xi_1^{-1} + \xi_2^{-1} + \xi_3^{-1} = 0.$$

The locus and envelope equations of the singular conics are respectively

$$\sqrt{x_1} + \sqrt{x_2} + \sqrt{x_3} = 0 = x_0, \quad \text{and} \quad \xi_1^{-1} + \xi_2^{-1} + \xi_3^{-1} = 0; \text{ etc.,}$$

and the quadric containing them is

$$x_0^2 + x_1^2 + x_2^2 + x_3^2 - 2x_2 x_3 - 2x_3 x_1 - 2x_1 x_2 - 2x_0 x_1 - 2x_0 x_2 - 2x_0 x_3 = 0.$$

The locus and envelope equations of the nodal lines are respectively

$$x_0 = x_3, \quad x_1 = x_2, \quad \text{and} \quad \xi_0 + \xi_3 = \xi_1 + \xi_2; \text{ etc.}$$

The mapping of (k) on π is according to the formulae

$$x_0 = (u_1 + u_2 + u_3)^2, \quad x_1 = (-u_1 + u_2 + u_3)^2, \quad x_2 = (u_1 - u_2 + u_3)^2,$$
$$x_3 = (u_1 + u_2 - u_3)^2;$$
$$u_1^{-1} = x_0 + x_1 - x_2 - x_3, \quad u_2^{-1} = x_0 - x_1 + x_2 - x_3, \quad u_3^{-1} = x_0 - x_1 - x_2 + x_3.$$

4. With the above notations, the relations between the tropes and the principal tetrahedron planes are given by the orthogonal substitutions:

$$\begin{aligned}
2x_0 &= w + x + y + z, & 2w &= x_0 + x_1 + x_2 + x_3, \\
2x_1 &= w + x - y - z, & 2x &= x_0 + x_1 - x_2 - x_3, \\
2x_2 &= w - x + y - z, & 2y &= x_0 - x_1 + x_2 - x_3, \\
2x_3 &= w - x - y + z, & 2z &= x_0 - x_1 - x_2 + x_3.
\end{aligned}$$

Complementary quartic curves on (k)

14·4. A quadric meets (k) in a curve C^8 which is mapped on π by a curve C^4. The quadric meets each nodal line in two points, and the curve C^4 meets each side of the self-polar triangle of the basal range in two pairs of points of the basal involution. Conversely, any C^4 which satisfies this condition with regard to the exceptional points represents a quadric section of (k).

The quadric section of (k) may degenerate into two quartic curves C_1, C_1' and then the corresponding curve C^4 on π consists of two conics. The quartic curves C_1, C_1' are therefore rational and lie $(3, 1)$ and $(1, 3)$ on the quadric. As only one quadric can be drawn through C_1 or C_1', they are termed *complementary* on (k).

If the conic representing C_1 meets a side of the polar triangle in A and B, then the conic representing C_1' meets it in A' and B', where AA', BB' are pairs in the basal involution on that side. These conics are reciprocal with respect to the conic $u_1^2 + u_2^2 + u_3^2 = 0$, which represents the section of (k) by the principal plane.

The conics reciprocal to those of the basal range form a pencil passing through four points constructed from the quadrilateral according to the usual quadrangle-quadrilateral construction. Hence the curves complementary to the asymptotic curves pass through four symmetrical points of (k); and the four-tangent quadrics of (k) which pass through the four points cut out the system of asymptotic curves. These four points may be called the *principal points* on (k).

14·4,1. At each point of a nodal line of (k) there are two tangent planes. At two particular points of the line the tangent planes are coincident; the tangent planes at these points each pass through two of the principal points and form a plane pair of the above system of four-tangent quadrics.

2. Of the asymptotic curves on (k) there are two which are equianharmonic. They are complementary and are the only asymptotic curves which pass through the principal points. The principal points are the hessian points on each curve.

3. Any quartic curve on (k) which passes through the principal points meets the principal plane at four points given by the hessian of the quartic equation giving the principal points on it.

4. The principal points of (k) form a tetrahedron in perspective with the tetrahedron of tropes, the vertex and axis plane of perspective being the centre and the principal plane.

5. If the points of (k) are projected from the centre, upon the principal plane, and the transformation $X = x^{-1}$, $Y = y^{-1}$, $Z = z^{-1}$ is then used in the plane, the plane sections of (k) become conics, and the correspondence is equivalent to the mapping of (k) upon π.

6. From any point P of (k) there are four osculating planes of C_1 which do not contain osculants through P. The points of contact of these planes are in one plane through the centre and P. Conversely, the osculating planes at four points of C_1 in a plane through the centre meet at a point of (k).*

* Armenante, *Giorn. di Mat.* (1) 12 (1874) 252.

7. The conic osculants at four points of C_1 on a given plane meet at six points of (k). The tangent planes to (k) at these six points concur at the pole of the given plane with respect to (e).

8. The conic osculants of C_1 which touch the plane meeting C_1 in $(\alpha t)^4 = 0$ are osculants at the points given by the hessian

$$(\alpha\alpha')^2 (\alpha t)^2 (\alpha' t)^2 = 0.$$

9. The locus of the pole, with respect to the conic osculants of C_1, of a plane meeting C_1 in $(\alpha t)^4 = 0$ is a quartic curve given by

$$x_i = (\alpha\alpha_i)^2 (\alpha t)^2 (\alpha_i t)^2.$$

The pole lies in the plane meeting C_1 in $(\beta t)^4 = 0$ if the corresponding osculant is that at one of the points given by $(\alpha\beta)^2 (\alpha t)^2 (\beta t)^2 = 0$. The points of contact of the four osculants so obtained are coplanar if the two given planes are conjugate with respect to the quadric (f).

10. For any of the surfaces K^*, K or (k), the locus of the pole of a given prime with respect to the conics on the surface is another surface of the same kind as the original.*

Particular quartic curves C_1

15·1. The curves which require special discussion are those for which the f-quartic either (i) admits of further linear transformations into itself so that the curve admits of other self-collineations than the three symmetries and thus has further projective features, or (ii) has not four distinct roots so that the analysis given above does not directly apply.

The first case can only occur if the six cross-ratios of the four roots are not distinct, and so there are just two possibilities, namely $i = 0$ and $j = 0$.

The possibilities in the second case are that

(a) two roots are equal $(i^3 - 27j^2 = 0)$,

(b) three roots are equal $(i = j = 0)$,

(c) there are two pairs of equal roots $((gT)^6 = 0)$.

When all four roots are equal, C_1 is a normal cubic curve.

It is clear that the curves arising under any one of the above headings are projectively equivalent; the projective classification is therefore completed.† Each type is obtained by projecting the normal curve C from a suitable vertex.

---

* Lie's theorem in Bertini, *Geometria proiettiva* (1923) 430.

† Reality distinctions lead to further discrimination: v. Telling, *Proc. Camb. Phil. Soc.* 29 (1933) 470.

The equianharmonic curve C_1

15·2. This curve is the projection of the normal curve C from a general point F of the quadric I, and its flexes are equianharmonic on the curve. The point F may be taken (4·2, 3) as (1, 0, m, 0, 1), where $m = \sqrt{-3}$, the f-quartic then being $t^4 + 6mt^2 + 1 = 0$, so that the hessian points are given by $t^4 - 6mt^2 + 1 = 0$, and the equations of the curve referred to the principal tetrahedron are as in 9·22 with $\mu = \sqrt{-3}$. The equation of the quadric (h) is $x^2 + \epsilon^2 y^2 - \epsilon z^2 = w^2$ where $\epsilon^2 + \epsilon + 1 = 0$, and of the quadric ($e$) is $\xi^2 + \eta^2 - \zeta^2 = 0$; the latter is therefore a conic on the principal plane.

In [4] the polar prime Φ of F is the tangent prime of I and thus incident with F. Hence the figures of projection and section have to be considered separately.

The prime Φ being tangent to I meets it in a quadric cone: the tangents, osculating planes, and osculating primes of C meet Φ in the points, tangents and osculating planes of E^6, which lies on the cone and meets each generator three times. The cone therefore projects from F into the conic (e) which replaces the equianharmonic quadric of the general case; *the flexes of C_1 form an equianharmonic set on the conic* (e) and the flex planes touch (e) at the flexes, while the curve E^6 is projected into the conic (e) triply. Thus *three tangents of C_1 pass through each point of* (e) (13·1), the tangent of (e) at the point being a triaxis of C_1; each triad of points of contact of the tangents with C_1 determines the h-polar plane of their point of concurrency, and this plane thus envelopes the quadric cone (q) which is the h-reciprocal of (e). The cone (q) has its vertex at the centre and touches C_1 at the hessian points, its equation being $x^2 + \epsilon y^2 - \epsilon^2 z^2 = 0$ (9·5).

In the section by Φ, the osculating planes at the four cusps of E^6 are concurrent in F, and the tangents, being traces of the tangent planes of K, are the tangent trisecants of $C^\star \equiv (K, \Phi)$. Hence both E^6 and $C^\star$ are equianharmonic curves with their centres at F, and the quadric cone (I, Φ) is related to $C^\star$ as (q) is to C_1. The nodal curve of E^6 is $C^\star$ and the bitangent developable of $C^\star$ is the osculating developable of E^6. Thus it follows that the bitangent developable of C_1 is equianharmonic; its edge lies on

(q) and is a sextic curve with cusps at the hessian points, and cusp tangents which are the tangent trisecants of C_1.

The nodal curve of C_1 is clearly the conic (e) counted three times. The steinerian points on C_1 are at the flexes, and the hessian points are the principal points (14·4) on the surface (k) associated with the curve.

The roots of the equianharmonic f-quartic are transformed into themselves by a linear transformation s which leaves one of them fixed and cyclically changes the other three. When this is combined with the axial group the resulting twelve operations form a tetrahedral group; the eight new operations are each of the same nature as s and are inverse in pairs. *There is thus a tetrahedral group of collineations which transforms the curve C_1 into itself*, formed by combining the symmetries with a cyclic collineation of order three inducing on C_1 a projectivity s which interchanges cyclically three flexes and three corresponding hessian points whilst leaving the fourth flex and hessian point fixed. Now the centre and a hessian point are in line with a vertex of the tetrahedron of flex planes, and since these three points are fixed in this cyclic collineation, this line must be a line of double points. Hence the collineation may be described—regarding the principal plane as the plane at infinity and the quadric (h) as a sphere—as a twist through $\frac{2}{3}\pi$ about this line; the tetrahedron formed by the hessian points and the homothetic tetrahedron of the flex planes are then regular, and the principal axes bisect their opposite edges: the transformations of the curve then appear as those of a regular tetrahedron into itself. But with the projection assumed above C_1 is imaginary; for the real curve the collineation may be described as a generalised (non-euclidean) twist with the conic $\xi^2 + \epsilon\eta^2 - \epsilon^2\zeta^2 = 0$ as absolute conic.

15·2,1. This tetrahedral group of collineations transforms into itself the surface (k) associated with C_1, as also the curve C_1' which is complementary to C_1 on (k).

2. Of the asymptotic curves on (k) there are three which are harmonic. A cyclic collineation transforming C_1 into itself transforms these three curves into one another.

3. On the map of (k) the curves C_1 and C_1' may be represented by conics $x^2+y^2-z^2=0$, $x^2+\epsilon y^2-\epsilon^2 z^2=0$, while $\xi^2+\epsilon\eta^2-\epsilon^2\zeta^2=0$ is the reciprocator of any pair of conics representing complementary curves; these are the equations of (e) and of the sections of (q) and (h) by $w=0$. The basal quadrilateral of the map is the section of the flex planes.

4. The curves C_1 and C_1' have the same conic (e) and cone (q). They are transformed into one another by any one of six harmonic perspections which, combined in pairs, give the three symmetries. By either transformation of one of these pairs one of the harmonic curves is transformed into itself while the other two are transformed into one another.

5. The surface (k) is transformed into itself by an octahedral group of collineations isomorphic with the group of twenty-four permutations of the four lines of the basal quadrilateral of the map.

The harmonic curve C_1

15·3. This is the projection of the normal curve C from a general point F of its chordal J: *thus C_1, whose flexes form a harmonic set upon the curve, has a node.* The point F being taken $(4\cdot2,3)$ as $(1, 0, 0, 0, 1)$, the f-quartic is $t^4+1=0$; the hessian points are at the node $(t=0, \infty)$. The equations of the curve referred to the principal tetrahedron are as in 9·22 with $\mu=\infty$; one principal axis is the intersection of the osculating planes at the node $(y\pm z=0)$ and will be called the *main axis* of C_1 and the opposite edge the main directrix; the principal plane is bitangent at the node. *The curve lies on a pencil of quadrics touching the principal plane at the node.* In particular the quadric of "trisecants" is a cone with its vertex at the node, its equation being

$$(h): \qquad y^2-z^2=\tfrac{4}{9}w^2.$$

The equianharmonic quadric (e) is given by

$$(e): \qquad \xi^2+\eta^2-\zeta^2=3\omega^2.$$

In the pencil of quadrics there are two further cones, vertices P and P'. A plane touching either of these cones is bitangent to C_1 and thus its points of contact are related by $t_1^2 t_2^2+1=0$, which breaks up into the two involutions $t_1 t_2\pm i=0$ corresponding to the two cones; the joins of t_1, t_2 are the generators of the cones which together form the bitangent developable of C_1.

The distinctive property of a harmonic quartic is that it remains unaltered under an involution having double points at one pair of roots of the quartic: $t^4 + 1 = 0$ is unchanged under the involutions $t_1 t_2 \pm i = 0$. These involutions are induced on C in [4] by harmonic inversions with director lines p and p' through F, and director planes ϖ and ϖ' in Φ. On projection from F upon Φ the lines p and p' become the vertices P and P', and the planes ϖ and ϖ' become the axis planes of two harmonic perspections transforming C_1 into itself, the joins of the corresponding points on C_1 forming the cones above. The axis planes ϖ and ϖ' must pass through the main axis; P and P', being their poles with respect to (e), lie on the main directrix and are incident with ϖ' and ϖ respectively. The prime Φ touches the surface K and meets it in two conics ($C^\star$ on ϖ and ϖ') which pass through the centre and are the e-reciprocals of the two cones.

The eight projectivities formed by combining $tt' = i$ with the principal involutions form a dihedral group. One of the four new operations is the involution $tt' = -i$, and the other two $t' = \pm it$ are cyclic projectivities of order four. *Thus C_1 is transformed into itself by a dihedral group of eight collineations* consisting of the symmetries, the two harmonic perspections, and two cyclic collineations of order four.

15·3, 1. The cone with vertex P and the conic on ϖ touch the quadric (e) at two of the flexes.

2. The extremities of the two principal chords, with P and P', form the vertices of a complete quadrilateral.

3. Any plane through P and P' meets C_1 at the points of contact of a skew quadrilateral which circumscribes it (13·1).

4. The nodal curve breaks up into two cubic curves, in ϖ and ϖ' respectively; the node of C_1 is a cusp on each curve, the cusp tangent being the main directrix, while P' and P are the respective flexes.

5. A system of confocal paraboloids is inscribed in a rational quartic developable (with an edge C^6) which is harmonic. The main axis of the developable is the axis of the paraboloids and the main directrix is at right angles to it at infinity. The edge C^6 lies on a quadric whose principal axes are the main axis and two lines of the developable

which are also chords of C^6 (equally inclined to the principal planes), the centre being midway between the vertices of the focal conics; about these axes the developable is symmetrical. Any tangent to C^6 meets the principal planes of the system at points on the focal conic; there are skew quadrilaterals circumscribed to C^6 having alternate vertices on the two focal conics, while their alternate faces touch two cubic cylinders each with an axis perpendicular to a principal plane.

6. If any rational curve of order n in $[r]$ has a node, and the parameters at this node are taken as $0, \infty$, then one condition for n points to lie in a prime is $t_1 t_2 \ldots t_n = \text{constant}$.

If α, β are the parameters of the node, the condition is

$$(t_1 - \alpha) (t_2 - \alpha) \ldots (t_n - \alpha) = k \ (t_1 - \beta) (t_2 - \beta) \ldots (t_n - \beta),$$

and thus if there are s nodes with known parameters, s conditions for co-primality can be written down.

The quartic curve C_1 with one stationary tangent

15·4. This curve C_1 is the projection of C from a general point F of an osculating plane; the coordinates of F may be taken as $(0, 0, 1, 1, 0)$, the f-quartic then being $6t^2 + 4t = 0$. There are two distinct flexes and two distinct hessian points, the remaining two in each case being absorbed by the stationary tangent (at $t = \infty$). Suitable equations for the curve are

$$x_0 : x_1 : x_2 : x_3 = 1 : t : t^2 + t^3 : t^4.$$

If A is the point of contact of the osculating plane of C, the projection from F and the section by its polar prime Φ have each the tangent at A for stationary tangent and this line is a generator of the quadrics (e) and (h). The curve on the Φ section is of order five and has two distinct cusps; the Φ section of K also has the tangent at A as stationary tangent, so that the bitangent developable and the nodal curve (quintic) of C_1 have each this same line stationary at A.

The f-quartic with repeated root admits of only one linear transformation into itself, namely the involution which has $t = \infty$ for double point and interchanges the other two roots: $t + t' = -\frac{2}{3}$. *Thus the only collineation of C_1 into itself is a harmonic inversion* inducing this principal involution; it has an axis through A which is the one chord-axis of C_1. If the osculating plane at A is spoken of as at infinity the whole figure is symmetrical about this axis.

The harmonic envelope (k) related to C_1 is a Steiner surface with two nodal lines coincident in the stationary tangent, and with the triple point at A. The asymptotic curves are all quartics with the same stationary tangent at A and with flexes on the singular conics of the two tropes on the surface. They are mapped upon a plane into a system of conics touching three fixed lines, one of them at a fixed point.

The equation of the harmonic envelope (k) is $\sigma \xi_3 = \eta \xi_2^2$ (cf. 14·3, 1), where $\eta \equiv 3\xi_0 - \xi_1 + \frac{2}{9}\xi_2 = 0 = \xi_3$ gives the principal chord and $\sigma \equiv 8\xi_0 \xi_2 - 3\xi_1^2 = 0$ the singular conic which is osculant at the flex $t = 0$ and meets the stationary tangent. The singular conic on the repeated trope is the point-pair on the tangent where the other two tropes meet it.

The quartic curve C_1 with a cusp

15·5. This curve C_1 is obtained by projecting the curve C from a point F of a tangent, and naturally presents features which appear in both the harmonic and equianharmonic cases. The point F may be taken as $(0, 0, 0, 1, 0)$ (4·2, 3); the f-quartic then being $t = 0$, the condition of coplanarity is $\Sigma t_i = 0$, and the equations of C_1 may be written

$$x_0 : x_1 : x_2 : x_3 = 1 : t : t^2 : t^4,$$

by means of which results can be verified.

Let A be the point of contact of the tangent through F, and B that of the one osculating prime, through F, which does not contain the tangent. The projection from F and the section by the polar prime Φ, of the complete figure of C, give projectively equivalent curves, each being a quartic with one cusp and one flex, corresponding to A and B in the former, and to B and A in the latter case. Hence *the curve C_1 is self-dual*, by which is meant that it is the dual of its own developable. On considering the curve C_1 of the Φ section it appears that *its nodal curve is a conic (e)* (section of K) and *its bitangent developable is a quadric cone (e′)* (section of I), while (e) and (e') correspond in the duality of the curve. The conic (e) touches C_1 at the cusp and touches the flex plane at the flex; the cone (e') touches the flex plane along the flex tangent and touches the cusp osculating plane.

The curve C_1 is the base of a pencil of quadrics having stationary contact at the cusp; there are only two cones in this pencil, the cone (e') and a cone (h) of "trisecants" with its vertex at the cusp. Dually the osculating developable of C_1 is the base of a range of quadrics having stationary contact at the flex; the range contains the conic (e) and one other conic, (h'), which lies on the flex plane.

The f-quartic has a triple root $(t = \infty)$ corresponding to the cusp and a simple root $(t = 0)$ corresponding to the flex. An infinite number of projectivities $(t' = ct)$ transform these points into themselves; *thus an infinite number of collineations, $y_0 = x_0$, $y_1 = cx_1$, $y_2 = c^2x_2$, $y_3 = c^4x_3$, transform C_1 into itself.* Of these collineations only one is involutory $(c = -1)$; it is a harmonic perspective with vertex at the point $(0, 1, 0, 0)$ and axis plane $x_1 = 0$. This perspective induces the involution $t + t' = 0$ on C_1; the points t, t' are the points of contact of the bitangent planes of C_1, their joins are the generators of (e'), while the tangents at them meet in the points of (e) on $x_1 = 0$.

There are, however, involutions $tt' = a$ which transform the cusp and flex into the flex and cusp; they leave the bitangent relation $t + t' = 0$ unaltered, and are induced by quadric polarities which change the point t into the osculating plane at t', namely

$$\xi_0 = 3a^4x_0, \quad \xi_1 = -8a^3x_1, \quad \xi_2 = 6a^2x_2, \quad \xi_3 = -x_3.$$

Any one of these polarities reciprocates the curve and its developable into one another. Thus C_1 is transformed into itself by a continuous group of ∞^1 collineations, while this group is extended by any one of the polarities to form the group of all collineations and reciprocities of the curve into itself.

15·5,1. The curve C_1 is of rank five; it is the only curve in [3] of this rank.

2. The osculating planes through any point of the plane of (e) have their points of contact on a plane through the vertex of (e'); these are the only concurrent osculating planes with coplanar points of contact.

3. The harmonic envelope (k) of C_1 has one trope (the flex plane) and one nodal line (the cusp tangent) with a triple point at the cusp. The equation of the envelope is $\xi_2 \rho = \xi_3 \tau$, where $3\rho \equiv 12\xi_0 \xi_3 + \xi_2^2 = 0$ and $2\tau \equiv 32\xi_0 \xi_2 - 9\xi_1^2 = 0$ are the conics (e) and (h') which are touched by all the osculating planes of C_1, and $\xi_2 = \xi_4 = 0$ is the cusp tangent The locus equation of (k) is $x_4 x_0^3 = \sigma^2$, where

$$\sigma \equiv 3x_0 x_2 - 2x_1^2 = 0 = x_4$$

is the singular conic.

The asymptotic lines on (k) are cuspidal quartics with flexes on the singular conic; they are mapped into a range of conics osculating at a fixed point and touching a fixed line.

4. For any rational curve of order n in $[r]$ with a cusp, one condition that n points are in a prime may be taken as

$$\frac{1}{t_1 - \alpha} + \frac{1}{t_2 - \alpha} + \dots + \frac{1}{t_n - \alpha} = \text{constant},$$

where α is the parameter of the cusp.

The quartic curve C_1 with two stationary tangents

15·6. This curve C_1 is the projection of C from a point F —$(0, 0, 1, 0, 0)$—on the surface K: through F pass two planes osculating C at points A and B. The f-quartic is $t^2 = 0$ and is identical with the hessian equation, while the equations of C_1 may be written

$$x_0 : x_1 : x_2 : x_3 = 1 : t : t^3 : t^4.$$

The polar prime Φ of F is bitangent to C at A and B, and its section of the figure of C gives a quartic curve with stationary tangents at A and B. *Thus C_1 is self-dual.* Moreover, *the tangents of C_1 belong to a linear complex* since one of the linear complexes containing C has its centre at F (7·1); and in fact, as C_1 is of rank six, the linear complex in [3] determined by the two stationary tangents and any other three tangents must contain every tangent of the curve. Consequently the point t and the osculating plane at t on C_1 correspond in a null polarity; the four points of contact of the osculating planes through any point lie on the null plane of the point, the equations of the polarity being

$$\xi_0 = x_3, \quad \xi_1 = -2x_2, \quad \xi_2 = 2x_1, \quad \xi_3 = -x_0.$$

The prime Φ meets the surface K in a quartic curve with stationary tangents at A and B, so that both the edge of the bitangent developable and the nodal curve are quartic curves with these same stationary tangents at A and B.

The f-quartic is transformed into itself by any projectivities of the form $t' = ct$, or $tt' = a$. *Thus there are two ∞^1 systems of collineations of C_1 into itself.* Of the first system only one is involutory ($c = -1$), given by $y_0 = x_0$, $y_1 = -x_1$, $y_2 = -x_2$, $y_3 = x_3$; it is a harmonic inversion with axes $x_1 = x_2 = 0$ (AB) and $x_0 = x_3 = 0$. When combined with the null polarity above, it gives a second null polarity which transforms each point t into the plane through t and osculating C_1 at $-t$. Thus the lines joining t and $-t$ are ∞^1 *principal chords* (or chord-axes) of C_1; they are the projection of the conics of K which pass through F and their locus is *the surface* (k), *a cubic scroll with AB as nodal line* (6·22), in this case both the harmonic envelope and harmonic locus of C_1.

The collineations of the second system are all involutory; each is a harmonic inversion with a principal chord of C_1 as one of its axes. On combination with the null polarity there are ∞^1 quadric polarities any one of which reciprocates C_1 into its developable.

15·6,1. The quadrics (h) and (e) are respectively the equianharmonic locus and envelope of C_1.

2. The second nul polarity transforms the nodal curve of C_1 into its bitangent developable. There are ∞^1 quadrics any one of which reciprocates C_1 into its bitangent developable.

3. The asymptotic curves on (k) are its intersections with the pencil of quadrics $x_1 x_3 = \lambda x_0 x_4$; they have stationary tangents at A and B, and map into a pencil of bitangent conics on the plane.

4. The Φ section of the primal J is the surface (k) associated with the quartic curve which is the Φ section of the figure of C. On this surface lies the nodal curve of C_1 and the planes of the bitangent developable to C_1 touch it.

5. The curve on Ω in [5] representing the tangents of C_1 is a C^6 in [4] with two cusps, which projects from one point into a C^3 in [3] representing the chord-axes of C_1 (11·1).

NOTE ON INVOLUTIONS ON C

16·1. It will have been observed from the foregoing that the geometry of C might be regarded as a study of the simpler involutions (g_2^1, g_3^1, g_4^1) on the curve. In this note we add a few examples of constructions arising from some further involutions.

The involutions g_4^r are clearly at home on C: they are represented by the lines, planes and primes of the space. Also a g_4^1 and g_4^2 are conjugate when they are represented by polar elements, one by a line and the other by primes through this line. In that case the double points of g_4^1 are the triple points of g_4^2, and any pair of points on C which is apolar for a member of the g_4^1 is *neutral* for g_4^2, that is to say, it imposes only one condition that a set of the g_4^2 should contain it. Involutions g_n^r, $n \neq 4$, are most naturally studied on the normal curve C^n in $[n]$. When transferred to C the constructions arising are more elaborate and are best carried out with an eye on the behaviour of the involutions on C^n. It may be that the actual transformation from $[n]$ to $[4]$ resulting in the transference of the involution may yield useful results.* Involutions of lower order (g_2^1, g_3^1) have already been dealt with, the case of conjugate g_3^1 occurring in 9·5; there remain involutions of order higher than four.

16·2. Each set of an involution g_5^r on C forms the corners of a simplex and the faces of these simplexes envelope a manifold of dimensions r. On examining the number of faces which pass through r arbitrary points of C we find that the class of the manifold is $4 - r$. Dually, the osculating primes at points of a set form a simplex and the vertices of this simplex trace out a quartic curve C', a cubic scroll F, a quadric Q for $r = 1, 2, 3$ respectively; in what follows we speak in terms of this dual construction.

Residual to a fixed point t_0 in the g_5^3 is a g_4^2; hence the vertex opposite to the osculating prime at t_0 lies on a plane of Q; Q is

<hr>

* For an example v. Telling, *Proc. Camb. Phil. Soc.* 28 (1932) 403.

therefore a cone. In fact there is one quartic which is apolar to all members of the pencil of quintics conjugate to the g_5^3: it belongs to all the residuals and so represents the vertex of Q. If the pencil is written as $(pt)^5 + \lambda(qt)^5 = 0$, the g_4^2 residual to t_0 is conjugate to $(pt)^4(pt_0) + \lambda(qt)^4(qt_0) = 0$; the plane of Q is given (by primes) on varying λ in this equation, and a plane of the complementary regulus on Q on varying t_0 with a fixed λ.

On the other hand, the sets of g_5^1 are apolar to a unique octic $(at)^8$ and C' is the reciprocal of C with respect to the quadric $R \equiv (ax)^4(ax')^4 = 0$ outpolar to C and meeting it at $(at)^8 = 0$ (3·3, 6). Then the conjugate g_5^3 is given by $(at)^5(a\lambda)^3 = 0$, and the residual of a point t_0 in the pencil by the quartic conjugate to $(at)^4(at_0)(a\lambda)^3 = 0$; these last are therefore equations of the point on C' which is the pole with respect to R of the osculating prime at t_0.

Similarly if t_0 is fixed in g_5^2, the residual g_4^1 gives a generating line of F traced out by the vertex of the simplex opposite to the osculating prime at t_0. The ∞^1 lines and ∞^2 conic planes of F are obtained as before from the first polars of the conjugate g_5^2. Thus in all cases *the first polars of the conjugate involutions present a two-rowed rectangular array indicating the generation of C', F, Q.* Should g_5^1 belong to g_5^2 and g_5^2 belong to g_5^3, then C' lies on F and F lies on Q, and all three present themselves in one array as in 2·33; Q is given by the vanishing of one determinant of the array, F as common to three Q's, and C' as common to six.

If a g_5^3 belongs to a g_5^4, its Q-cone passes through a plane, namely that plane given by primes through all first polars of the unique quintic conjugate to the g_5^4.

16·2,1. The residual g_4^2 of t_0 in g_5^3 appears on the cubic osculant of t_0. The osculating planes at a set form a tetrahedron whose vertices trace out a quadric which is the section of Q by the osculating prime. Similarly the section of F is the reciprocal of the developable of the osculant, with respect to a certain quadric.

2. The point representing $(xt)^4 = 0$ with reference to C' is the point representing $(ax)^4(at)^4 = 0$ with reference to C.

3. The eight quadruple points of $(at)^5(a\lambda)^3 = 0$ are the points

where Q meets C and the osculating primes thereat touch R. There are twelve osculating planes and twelve tangents touching R.

There are fourteen outpolar quadrics which touch eight given osculating primes of C, and fourteen cones Q which pass through eight points of C.

4. There are nine tangents of C which are chords of F (triple points of g_5^2), and six axis planes which are conic planes of F (neutral pairs of g_5^2). These last meet any osculating prime in six lines which are one-half of a double-six.

5. Contained in a g_5^2 there is one g_5^1 which has more than one quartic apolar to all its sets (directrix of F).

6. Every quadric through F is a cone with its vertex on F. If one simplex circumscribing C has its vertices on a quadric, then there is a quartic curve C' on this quadric.

7. Hexahedra formed by the osculating primes at sets of a g_6^r on C have their vertices on a locus of dimensions r, and order 10, 6, 3 for $r = 1, 2, 3$, respectively.* Their generation is indicated by a three-rowed array formed by second (mixed) polars of the conjugate involutions.

If a g_6^3 is contained in a g_6^4, the cubic primal passes through a cubic surface on which lie the six nodes of the primal; if it is contained in a g_6^5, it passes through a definite line. This is the continued interpretation of the rectangular array.

8. In the case of g_6^3, the line edges of the hexahedra which lie on the cubic primal form a scroll of order twelve; four of these lines pass through each node, and six appear on each osculating prime as one-half of a double-six. Each line is tetrasecant to the sextic surface arising from any g_6^2 contained in the g_6^3.

16·3. The sextic involutions have already presented themselves in the constructions relating to complexes apolar to C (7·2, 5) and in the paragraph above (16·2, 7). The following method, virtually due to Stahl, makes direct use of the normal curve in [6] for the case of g_6^1. We first state the analogous construction for a g_4^1 on a conic.

In [4] the primes through a general plane, σ, meet C in a g_4^1; the faces of the tetrahedra of this series meet σ in the lines of quadrilaterals circumscribed to a conic γ, touching it at sets of a g_4^1. If C is projected from any one of its chords onto σ, there are

* Meyer, *loc. cit.* 387.

triangles inscribed in the projected curve (conic c) which are circumscribed to γ. The vertices of the quadrilaterals touching γ at sets of a g_4^1 lie on a cubic curve (σ, J), opposite vertices being corresponding points on the curve. This curve is the jacobian of a net of conics reciprocating γ into the various conics c obtained by varying the chord vertex of projection.

In [6] the primes through a general [4], σ, cut the normal C^6 in a g_6^1. The faces of the hexads of the series meet σ in solids which are the faces of hexahedra circumscribed to a normal C in σ. The involution is thus transferred to C and is a general g_6^1 on C. If the C^6 is projected onto σ from any one of its chords, there are simplexes inscribed in the projected curve, C', which are circumscribed to C, and thus these two curves are reciprocal with respect to a quadric R.

These propositions are readily proved from the geometry of the figure; in algebra they may be put thus: All the sets of a g_6^1 are apolar to a unique form $(at)^{10}$. The conjugate g_6^4 is then $(at)^6(ax)^4 = 0$; this equation represents each point of σ, while $(xt)^4 = 0$ represents the same point with reference to a certain curve C in σ. If $(pt)^5 = 0$ gives the five points of C^6 on a [4] meeting σ in a solid, $(ap)^5(ax)^4(aT) = 0$ holds for ∞^3 points (x). Hence $(ap)^5(at)^5$ is a complete fifth power $(t_1 t)^5$ and (x), satisfying $(t_1 x)^4(t_1 T) = 0$, is a point of a solid osculating C.

Again, from the argument of p. 49 it appears that the point (x), referred to the projected curve C', represents the equation of second polars $(at)^4(ax)^4(ac)^2 = 0$, where $(ct)^2 = 0$ gives the ends of the chord-vertex of projection, and thus the reciprocating quadric R is $(ax)^4(ax')^4(ac)^2 = 0$.

The hexahedra osculating at sets of g_6^1 on C are polar for each of the net of quadrics R.* The vertices of the hexahedra are the vertices of cones of this net, and lie on a curve of order ten, the jacobian curve of the net, and the trace of the line-chordal of C^6. The planes of the hexahedra, opposite to the vertices, touch the K-surface of C at the points of another curve of the tenth order.

---

* The quadrics of a general net in [4] have no common polar hexahedron, and the jacobian curve of the net has only a finite number of trisecants.

The lines which are edges of the hexahedra form opposite pairs of a scroll which lies on a quartic primal, an analogue of the symmetroid in [3], and the trace of the plane-chordal of C^6. Each of the lines is trisecant to the jacobian curve, and each of the planes meets the quartic primal in a quadrilateral whose vertices lie on that curve. (Cf. Stahl, *J. f. Math.* 104 (1887), 38, Conner, *Amer. J.* 37 (1915), 29, White, *Proc. Camb. Phil. Soc.* 23 (1927), 882.)

16·3,1. The sextic surface (cubic primal) arising from hexahedra osculating at sets of a g_6^2 (g_6^3) containing the g_6^1 is the locus of the pole (polar line) of a solid (plane) in the polarities $(ax)^4 (ay)^4 (ac)^2 = 0$.

Printed in the United States
By Bookmasters